2012 Edition

THE ESSENTIAL INVENTOR'S GUIDE

Step-by-step methods to successfully evaluate, patent, and market your invention on a budget

Robert K. Masse, Ph.D.

www.InventionPatentInformation.com

Copyright © 2009, 2010, 2011, 2012 by Robert K. Masse. All rights reserved.

ISBN: 1-4392-6190-3

CONTENTS

About This Document   v

1 Patent Basics   1

   1.1 Why Does the Government Grant Patents?   1
   1.2 The USPTO   2
   1.3 Who Enforces Patents?   2
   1.4 Types of Patents Granted by the USPTO   3
   1.5 International Patent Applications   5
   1.6 Expectations, Great or Otherwise   5
   1.7 Timing   8
   1.8 Getting Preliminary Protection   10
   1.9 Promoting Your Invention   11

2 To Patent or Not to Patent?   13

   2.1 Should I Patent My Idea?   13
   2.2 Can My Invention Be Patented?   14
   2.3 Is My Invention Lame?   17
   2.4 Is my Invention Worth Patenting?   18

3 Conducting a Patent Search   22

   3.1 Let's Find Out If You're the First   22
   3.2 Patent Search   22
   3.3 Part I – Keyword Search   24
   3.4 Part II – US Classification Code Search   27
   3.5 Final Tips Before you Start Your Search   29
   3.6 OK, Let's Get Started.   31
   3.7 When You're Finished   32

4 Making It Real   34

   4.1 Prototyping   34
   4.2 Design – More Important Than You May Think   37
   4.3 Getting Down to It   38

5 Starting the Application   41

   5.1 What You Will Need   41
   5.2 General Requirements to Apply for a Patent   44
   5.3 The Parts of a Patent   45

6 Drawings   47

   6.1 Drawing Rule Enforcement   47
   6.2 Learning by Example   48
   6.3 Detailed Listing of Drawing Rules   53
   6.4 Drawing Quality   65
   6.5 Let's Get to Work   66

7 Starting the Specification — 68

   7.1 Let's Start Writing — 68
   7.2 Background of the Invention — 71
   7.3 Brief Summary of the Invention — 76
   7.4 Brief Description of the Several Views of the Drawing — 78

8 Detailed Description of the Invention — 81

   8.1 Detailed Description of the Invention — 81
   8.2 Part Numbering — 82
   8.3 Vocabulary — 83
   8.4 Let's Write — 85
   8.5 Adding Reference Numbers to the Text — 89
   8.6 Adding Reference Numbers to Drawings — 92
   8.7 Closing Statement — 94

9 Claims — 96

   9.1 Claim Basics — 96
   9.2 The Structure of Claims — 100
   9.3 Qualifying Language — 105
   9.4 Terminology — 107
   9.5 Quick Quiz — 118
   9.6 Other Common Mistakes — 119
   9.7 Method vs. Apparatus Claims — 120
   9.8 Test Yourself — 122
   9.9 Claims by Organized Design — 126
   9.10 A Final Word of Advice — 128

10 Finishing Up — 131

   10.1 Abstract of the Disclosure — 131
   10.2 Completing Your Patent Application — 132
   10.3 Submitting Your Application — 143
   10.4 Other Resources — 148

11 Licensing Your Invention — 150

   11.1 Selecting a Licensee — 150
   11.2 Finding Prospective Licensees — 155
   11.3 Making Contact — 156
   11.4 The Solicitation — 159
   11.5 Oral Briefing — 168
   11.6 Non-disclosure Agreements — 172
   11.7 Negotiations — 172
   11.8 Exclusive vs. Non-exclusive Licensing — 175
   11.9 Compensation — 177
   11.10 Infringement — 180
   11.11 Off You Go. — 182

Index — 184

ABOUT THIS DOCUMENT

Dear Reader, presumably you have purchased *The Essential Inventor's Guide* because you have an idea that you are interested in patenting and marketing, and expect this guide to facilitate the achievement of those goals. This is what you have paid for, and this guide will provide you a step-by-step means to exactly those ends. Additionally, I will, where I have information that might be worth sharing, provide advice to the prospective patenter. Keep in mind, this is only advice, and it is provided with absolutely no guarantees, but the best of intentions. If you follow this advice, you do so at your own discretion; if you disregard it, don't say you weren't warned.

Who is this guide for?

If you're an inventor, that would be you. Many readers may as well skip this paragraph – you know well that inventorship is in your blood. Others, of course, will be less sure. If you have an invention, when should you consider doing something about it? Are you employed or unemployed? Do you love or hate your job? Are you married? Kids? Retired? Regardless of your circumstances, the answer is the same. *As soon as possible*. It's not at all about being in a hurry. It's about being proactive. He who is slow but determined will finish, but he who is complacent will never start. Will you finish? You can, although that may or may not be with the success you first envision. In the chapters that follow, we will go on at some length on everything that must come together to take an invention from that initial spark of inspiration all the way to profitable product; and it is a considerable journey. The honest truth is that few inventions go the distance, but there is more to the picture than just the few that strike it rich to make such enterprise worthwhile…and, of course, there are the ones that really do. Regardless, never let complacency be part of the equation.

To my youngest readers:

If you're not yet an adult (and perhaps not even close), do not let that be a concern. You are very welcome here, and this guide is every bit as much (and perhaps especially) for you. Not so long ago that I was also a youth with bigger-than-average plans (admittedly it was less recently than I usually prefer to dwell upon), and so you hold a special place in my heart. If the world is about anyone, it's truly you.

I do very much encourage you to enlist, and hope that you have available to you, a parent or other adult who is supportive of your project; but nevertheless, you are never too young to move from dreaming to doing…and, regardless of how this invention turns out, you'll find having it on record that you filed your first patent application so early in life a fine resumé item indeed.

Everything in this guide is based on the personal experiences of myself and others.

The information provided herein is based on research and the personal experience of the author, who was, not all that long ago, in exactly the same position in which you are now, and ever increasingly supplementary feedback from users. In the course of researching and writing my first and subsequent patents and continuations, the lack of a simple step-by-step guide was sorely noticed, and it occurred to me that I could much simplify the process, and greatly aid those that follow, through the creation of a learn-as-you-go workbook-style tool.

Use the information in this guide to your best advantage, but do so at your own risk.

I have already indicated that all that constitutes advice herein is provided without any guarantee. In this litigious day and age I'm certain that the reader will understand that. In fact, I must apply the same disclaimer to all of the information contained herein – no guarantee is made or implied in regard to either the accuracy or usefulness of anything in this guide. I'll do my best to provide you the facts as I see them and some useful tips, but I can't afford to make promises any more than I can force your expectations to be reasonable. In short, what I will share with you is what has worked well for me (and others), and I'll help you avoid what has not.

Here's the inevitable disclaimer:

Any statements, views, recommendations, advice, etc. expressed herein are solely the opinions of the author. No warrantee of accuracy or that the information contained herein is factual is extended to the user or implied. The reader should understand the author makes no assertion that the opinions expressed herein are not worthless or detrimental. The user shall assume full responsibility for verifying that the information contained herein is factual at the time of use and holds the author harmless with regard to any direct or indirect damages that may result from the use of this guide. Use at your own risk.

Don't skip the Chapters 1 - 4 preliminaries.

So, that said, congratulations on your purchase. I believe that you will find the information in this guide very helpful and time-saving. There's a fair amount of up-front work before you even begin writing the patent – in fact, we won't even turn our attention specifically to starting your patent application until Chapter 5. The preceding chapters are important and should not be considered optional; you see, before you start writing, it's important that you first:

You won't want to invest time and money into a patent application until you…

- *Learn a little about patents and what's patentable*
- *Evaluate whether or not your invention's worth patenting*
- *Verify that your invention isn't already patented*
- *Plan for how you are going to reduce your invention to practice*

Without carefully performing these preliminaries, writing and filing a patent application is likely to prove a big waste of time and money.

Inventions are like relationships – after the initial excitement fades will come a decision as to whether you are willing to endure the rigors of making it real.

You may be excited about your invention (of course you are!), and feel you can't wait another minute to file a patent application before someone else beats you to the punch. That's natural, but to the veteran, all enthusiasm is held in reserve until after the above vital first steps are complete (and emotional investment is never part of the process at all). An excellent idea is but the beginning. One that withstands the rigors of scrutiny, has been verified to be original, and is ready to market – well that's really something to be excited about…and so (following some basics) we'll first set out to qualify your invention in those respects. Start at the beginning and use your initial enthusiasm to charge through the first chapters – quickly if you can, but thoroughly above all.

You will need only basic word processing skills.

Once those preliminaries are complete, Chapters 5-10 will step you through the task of creating a patent application. Every effort has been made to simplify the process to the greatest extent possible without sacrificing quality of the end product. Likewise, as the user is assumed to possess only basic word processing skills, advanced techniques are employed only in very limited cases where they really save much time and frustration. Where advantageous (a particularly good example being the integration of drawing and text reference numbers), this guide will teach such techniques using Microsoft Word for illustration (as this will be of greatest advantage to most users), but the same strategies can be employed using any modern word processing software.

I'll teach you the few advanced techniques you are better off learning than trying to do without.

And yes, I am aware there is a split infinitive on the cover of this book. I do that. Without remorse.

What is the purpose of the left-hand sidebar commentary?

The sidebar comments will help you remember key points.

The notes running along the left margin highlight key take-aways in the accompanying text. Devices such as this are becoming increasingly common in business proposals as their passive presence has been shown to improve retention of the subject matter, equally a constant goal of any instructive text such as this. Used more actively, they will provide you an excellent means to test your comprehension of the accompanying paragraph(s) if you stop to review them before continuing to the next page. If you have absorbed the material contained in the text, these comments should make sense to you, and you will find yourself able to expand upon them, mentally backfilling their context, basis, and relevant associated details. If you cannot, consider re-reading the text.

They will also save you time the second time you use this guide by facilitating quick review.

Later, they will serve you again by facilitating quick access to specific text to which you may wish to return as you work. Finally, if you put this guide down and return to it after a considerable lapse of time, you will find browsing the sidebar comments (and re-reading the details associated with those you don't fully understand) much more effective than skimming the text in its entirety. As I said, this guide was designed for busy people, as true inventors are inclined to be.

Action Steps

You will perform twenty action steps to evaluate, prepare your invention, and compose/compile your patent application.

Throughout this guide, when it's time to perform any of the tasks you need to evaluate your invention and compose the patent application, you will find clearly denoted "Action Steps". You should carefully read all of the text leading up to each Action Step before attempting to perform it. Most Action Steps may be worked directly in the companion electronic *Patent Application Workbook*.

You may download the *Patent Application Workbook* here:

http://www.inventionpatentinformation.com/freepatentworkbook

Most action steps are small, but not all.

Be forewarned that I will strongly recommend you construct a model or prototype as an integral part of preparing to market your invention (if you have not already done so).

The Action Steps vary in scope. Some of them will instruct you to complete pre-formatted sections of the patent application or filing forms. Others will direct you to supporting tasks on worksheets in the Application Workbook that will not be included in the final submittal (but which you should leave in place even after you are done for your own future reference), and perhaps the largest of all will direct you to fabricate a model or prototype. Once the twenty-three Action Steps of this guide are complete, the Application Workbook will contain your finished, ready-to-print patent application (technically you will need to replace a slipsheet with several simple forms from the USPTO's website, but I will step you through that when it's time).

1
Patent Basics

1.1 Why Does the Government Grant Patents?

Patents teach an invention in exchange for the temporary right to stop others from copying it.

If most people were asked the question "Why should one file a patent?" they would generally answer something to the effect of obtaining exclusive right to use an invention, and generally this is correct and the most common motivation. But, this is actually quite opposite to the real reason the government issues patents. Fundamentally, a patent teaches others how to do something, not restricts them from doing it. It is in the government's interest (and ours) to promote the sharing of information, including inventions, and the patent accomplishes exactly that. However, the average person isn't going to go to all of the trouble of disseminating knowledge of how to use/manufacture (or, more formally "practice" or "reduce to practice") their invention without getting something in return, and it is with the goal of creating such an incentive that the government provides temporary (twenty years from the application filing date, or seventeen years from the date the patent is issued, whichever is longer) exclusive usage rights to inventors in the form of a patent. The anarchists and libertarians among us may object to the use of the word "provide" in the last sentence, but, in this case, it is indisputable that the protection of an idea or "intellectual property" can only be obtained through governance of some kind. Whereas it may indeed be true that you might argue a fundamental right to practice your invention, it is only through some law issuing and enforcing agency that others may be restricted from copying you.

1.2 THE USPTO

The USPTO provides excellent resources for the small independent inventor.

In the United States this function is provided through the United States Patent and Trademark Office (USPTO) within the U.S. Department of Commerce. Now the U.S. Government in general is not very popular amongst the citizenry these days, and generally I'd say there are some pretty obvious and justified reasons for that. Keep in mind, however, that the U.S. government is an enormous conglomeration of agencies and other entities, and there are bright spots. The USPTO certainly gets a gold star from me. You will find their web-based resources to be excellent, which is why I emphasize that, though this guide will greatly accelerate your learning and the production of your first patent, you really could accomplish this same goal using exclusively the resources provided by the USPTO, as I did when I started out.

U.S. patents are among the most affordable in the world.

One of the pillars of this nation's (the U.S.) success is its adherence to respecting and promoting its citizens' property rights, including, and in particular, intellectual property, which has nurtured the development of one of the most (if not the most) free and expansive markets on the planet. It would be no exaggeration to say that much of the great wealth of the United States stems directly from the opportunities protected by, and the information sharing fostered by, the USPTO and its predecessors. Now, clearly, one might expect that the USPTO, as the de-facto gatekeeper to said most free and expansive market on the planet, enjoys rather high fees. Quite to the contrary, by comparative standards, the opposite is true. U.S. Patents are among the most affordable in the world. Here's something else to feel good about:

> Did you know that the USPTO has been entirely self-sufficient, operating only on patent application and maintenance fees since 1991?

The USPTO does not cost you one tax dollar. If only the average bus system held up the same standard – IF ONLY! Needless to say, I'm a big fan of the USPTO, and we'll leave it at that.

1.3 WHO ENFORCES PATENTS?

Who makes sure no one infringes your patent?

You do!

It is perhaps worth noting here, that the USPTO does not actually protect your invention or stop others from copying it – you do that, but the USPTO provides you the means. The USPTO grants you a license documenting your invention as your intellectual property. As with the rest of your property, it's up to you to take someone to court if they try to take it from you. For this reason, a patent is only as good as your ability to enforce it. Keep in mind, the patent examiner's is not the only scrutiny your patent may have to bear. If your invention really is the next big idea, others will want a piece of the action, and they will look for any excuse to limit the scope of your patent or call it invalid. If they do see a weakness, they won't take you to court. They'll just infringe and dare you to take them to court.

1.4 Types of Patents Granted by the USPTO

There are four types of patents administered by the USPTO:

Non-provisional Utility Patent

When you want to patent an invention, this is the type of patent of which you are thinking. This is by far the most common type of patent filed, and the one toward which you are currently working.

Provisional Utility Patent

A type of get-your-foot-in-the-door patent that's good for one year while you work on a non-provisional. The truth is, a provisional patent must still be fairly thorough to be worthwhile, so it won't save you much time, but you can get it at about 1/20th the price.

Design Patent

A patent protecting not an invention, but the visual ornamental characteristics embodied in, or applied to, an article of manufacture. The most obvious purpose of this type of patent is to prevent competitors from making exact or near exact knock-offs of a product.

Plant Patent

This is exactly what it sounds like, a patent that prevents others from copying a newly discovered or created breed of plant (the kind that grow).

A Word about Provisional Patents

Provisional patents can provide short-term protection while you work on a regular patent application.

A provisional patent's protection ends after one year…

but a regular patent usually can still be later filed unless you or someone else puts the invention "in use" first.

Provisional patents are a relatively new, introduced by the USPTO in June 1995 with the intention of giving inventors a low cost option to get started. A provisional application is very temporary and, if well drafted, conveys the same privileges as a regular utility application; but for only one year, by which time the applicant must either file a regular utility application, file a petition to convert the provisional application to a regular utility application, or lose all priority benefits of the original filing. Note that normally if the provisional application lapses, that does not prevent the filer from applying for a regular utility patent at a later date; however, if the invention goes "in use" or "on sale" during the twelve-month period of effectivity and the provisional patent is allowed to lapse, the right to file for a non-provisional patent may be forfeit. A regular patent that references a provisional is still good for the normal minimum of 20 years from the filing date or seventeen years from the date granted, so the provisional can actually extend protection for up to an extra year if filed before the provisional expires.

Provisional patents are effectively placeholders.

A good provisional patent will basically comprise a complete utility application sans the full filing fees.

Provisional patent applications are not reviewed by a patent examiner, and so there are not much in the way of rules regarding content. Generally, the USPTO recommends that a provisional patent be as complete as possible for the simple reason that only the parts of a later application that are directly supported by the provisional gain the provisional's filing date of priority. Unlike a regular utility patent, no claims are required for a provisional application...but, I caution you not to expect you've been thorough without including claims. Moreover, in the event you, for whatever reason, cannot file a follow-up application, you cannot petition to convert your provisional to a regular utility patent unless it contains claims, along with all of the other required components.

If the filing fee is a problem, a provisional application gives you a year to find a licensee to pay for the regular patent application.

Since you pretty much should complete all the same tasks for a provisional application as a non-provisional, the provisional, really has only two purposes. Firstly, you delay the filing costs (but, at the end of the day, you'll have paid an extra ~10% for the provisional) for up to almost one year (but don't cut it too close). If, in that time, you license your patent, the assignee will generally pick up the remaining costs, so, you could save quite a bit. Secondly, you can give yourself extra time to market your idea before the arrival of the first office action (required response to feedback from the USPTO). If you plan to license, it's highly desirable to do that before the patent issues, and even better if you can do so before any further communication with the USPTO is required, to take advantage of the licensee's legal resources to prosecute the application (to reduce your risk of making a mistake). I have not typically bothered with provisional applications, mainly because I'm not that strapped for cash, and the truth is, if you actively pursue licensing for a year and are unsuccessful, the odds of future success are rather small.

Provisional patents are sometimes used to ensnare inventors into expensive follow-on services.

Beware organizations encouraging you to file a quick provisional patent. They often do so hoping that months will pass and you will suddenly find yourself in desperate need of (their) professional services. Provisional patent applications are misused more often than not, and, if poorly constructed, can do more harm than good.

Here's a link to the USPTO's page on provisional patents, if you want to know more:

http://www.uspto.gov/web/offices/pac/provapp.htm

1.5 INTERNATIONAL PATENT APPLICATIONS

International patents just get the ball rolling.

You still must ultimately file for individual patents in all countries where you want protection.

Wouldn't it be great if you could just file one world-wide patent instead of having to file them in every country in which you want protection individually? Well, there is…sort of. Back in 1970, the World Intellectual Property Organization (WIPO), a then three-year-old agency of the United Nations, sponsored the creation of the Patent Cooperation Treaty, which established rules creating a universal patent application process for the more than 125 (currently) participating nations. That's as far is it went, however. If you want patents in any of those participating nations, you must, in a second phase, prosecute the application individually in each nation. That means even after you pay the international application fees, which are typically double the USPTO's, you must pay application and other fees for each of those nations where you want protection (but at least you get to pick and choose).

International patenting is very expensive.

Fortunately, you have up to a year after your U.S. filing to find a licensee to pay for any international patents you'll need.

Obviously, international patenting is, therefore, very costly, and something you'd like to delay until after you know that your patent is worth the expense (some countries do give you a discount on the filing fee if you already paid for an international application, but translation fees are also very common). Fortunately, you do not have to file the international application first. You can file in the U.S. first, after which you have up to twelve months to file the international application (that is, after filing either a provisional or non-provisional patent application). Once again, clearly it's in your best interest to license within the first year after filing, so that your licensee can foot the bill for much more expensive stuff like this.

Generally you should start with a domestic (U.S.) patent. International patenting is out of the scope of this book, and mentioned only because I don't want you to be spooked by the terminology when you see it around. If you'd like to know more about international patents and/or the PCT, the USPTO has a comprehensive link page:

> http://www.uspto.gov/patents/init_events/pct/index.jsp

Most of the links you'll see on this page actually take you to the WIPO's website, by the way.

1.6 EXPECTATIONS, GREAT OR OTHERWISE

Getting a patent will impress your friends.

Spending a lot of money on a patent will impress your friends that you are a fool.

So far you've shown good sense in that you have purchased *The Essential Inventor's Guide.* Good for you! You're right to want to keep costs low. Keep them very low; in fact, as low as possible is a good rule of thumb for most people. If you have substantially greater than average wealth, what's low for you may be proportionally higher than the mean, but what you spend on your patent should, regardless, always constitute absolutely negligible financial risk. The basic reason for this is that in pursuing a patent, or anything for that matter, your investment should be commensurate with your probable gains...and let's consider for a moment your probable gains.

Like any investment of time and money, one must weigh the cost against the probable gains.

Would you buy a lottery ticket for a dollar if the payout to the winner was one million dollars? You may say sure, but what if I ask you this – what is the fair price of a lottery ticket for a one million dollar jackpot? Hopefully now, if not before, you've realized that there is more to this than simply the price and the prize – you need to know your chances to win. If I told you that I was selling 100 million of those one dollar tickets for my $1M lottery, that wouldn't seem like a very good deal at all – I'm paying out $1M and pocketing $99M (sadly most state lotteries' payouts are indeed a pretty small fraction of total sales, which is why they are justifiably often labeled "taxes on the stupid").

NEVER spend a lot of money on a patent.

The blunt truth is that the statistical odds of making money on your patent are low.

Unfortunately, statistically speaking, your odds of making any money on your idea, patented or otherwise, are even smaller than that. Sorry, that's the truth. I know your idea is great, and you're very excited about it. Realize that so were all of the millions of patenters before you, and almost none of them made any money, but lots of them lost plenty. Consider that:

The honest truth is that you are probably too close to your idea to objectively evaluate it.

Ask people you can trust for their honest opinions regarding your invention's potential.

I strongly urge you to consider talking about your idea with several family members or friends, but you need people who you can implicitly trust and who will be completely, brutally honest with you. Listen to them. Remember all of the would-be inventors who auditioned on ABC's *American Inventor* back in 2007 and 2008? Most of their ideas were obviously downright bad (except maybe to George Foreman). So bad, in fact, that it was often hard to believe they couldn't see it. But they couldn't, and chances are you can't either. Having a bad idea doesn't in any way cast a shadow on your potential (I come up with them all too frequently). Smart people have bad ideas all the time; we call them smart because they also have good ones. What will meter your potential, however, is your ability to sort the good from the bad.

Pick a topic and do a search on the USPTO's patent database – here's the link:

http://patft.uspto.gov/netahtml/PTO/search-bool.html

You will be amazed by the amount of money that has been spent patenting goofy inventions.

Just start clicking and skimming. Do most of these seem like good ideas to you? (Please say no.) How about the ones that appear quite clever? Have you ever heard of them? If they were successful, wouldn't you have? (O.K., sometimes no, but keep scanning, and you'll get the point). Again, that's just the way it is, and we have to accept it.

Even if your idea is a good one, it will probably never make you a single dime.

Even many good inventions don't make any money.

Alas, even if you have a genuinely excellent idea, it probably will never be profitable for you. So many things can happen to bring a good idea to ruin, and so many things have to go just right to bring it to fruition. I remember a coworker who, with a buddy, had developed a technology that was brand new – not something no one else could have made, but the type of thing where he transferred existing technology to an entirely new application. (I'll respect his privacy by not explaining details.) Anyway, about $10k was spent, the device worked as planned, and a fortune seemed within reach...when suddenly something totally different, not nearly as slick, very low tech, producing an inferior but much cheaper end product, hit the market. That was the end of it. To say that no idea is fool-proof would be a gross understatement. The point is, that ideas really are a dime a dozen, and hardly justify the expenditure of thousands of dollars.

If wealth is your goal, there are many easier paths than inventorship, some of which are legal.

Do I mean to discourage you? Absolutely not. But you need to go into this with your eyes open. If you stay in control, you'll be proud of your invention and patent, even if it doesn't make you rich. Pursue your dreams, but live within your means (rhyme accidental, but not objectionable).

There are many paths to write a patent application. We can start a list:

---

1 The least expensive (if you get everything right the first time), is to sit down at your computer open a blank document, and start writing. That's how I started, and the USPTO has excellent resources for the independent inventor. As I previously indicated, you really can get most everything you need right on their website. For a government organization (and even as compared to many other organizations), they're top notch.

---

2 The second least expensive way to get a patent (and by relative comparison almost the same as 1 above) is to purchase this guide and follow the step-by-step instructions. There is quite a bit to know and get together, and this guide will simply save you a great deal of time and frustration.

BUT IT WON"T MAKE YOUR INVENTION SUCCESSFUL.

Only, a good idea, perseverance, time, timing, and luck can come together to do that.

A good patent attorney is valuable, but very expensive.

From here the options get significantly more expensive – you can engage a professional, either a professional patent writer or (even more expensive) a patent attorney. For the small independent inventor, this almost never makes sense. It's not that patent professionals have nothing to add, to the contrary, clearly since they do this full-time they have the potential to provide a great deal of experience-based insight (once they have some experience behind them). So, don't get me wrong, a professional, if you can sort a good one out of the pack, will absolutely make your patent more bullet-proof (if such a thing is truly possible).

8 PATENT BASICS

All things considered, patent attorneys aren't as overpriced as you may think, but that doesn't mean you can afford to have one draft your patent.

That extra help will be, however, expensive. Very expensive. One could almost go as far as to say, too expensive. Like all lawyers, it seems patent attorneys occupy the upper end of the food chain, but, the truth of the matter is, their compensation isn't quite as far out of family with other professionals with similar education as you might think. When people go to school that long, and put up with the hours junior lawyers endure, they expect a little payback once they've earned their stripes. So, right or wrong, the cost of having an attorney draft your patent is going to far exceed the level of investment justified by the miniscule chances you have of making any money on your idea.

Patent attorney or not, you'll still have to do most of the work; only you understand your invention.

Also, harbor no illusions, even if you engage a patent attorney, you'll still have to do most of the up-front work. A patent attorney may be an expert at law regarding intellectual property protection, but he doesn't know anything about your invention. It will still fall to you to draw it, describe it, etc. Basically, you'll still need to produce the real content of the drawings and at least a draft of the "DETAILED DESCRIPTION OF THE INVENTION", the most labor intensive portions of the patent application (but ironically not the most important – that would be the "CLAIMS"; don't worry if you don't know exactly what those are yet, we'll get into that later). Even if, up front, you do not feel comfortable taking your chances without professional help, first creating a draft patent application using this workbook, and then engaging professional services to merely review and edit will save you a ton of time and money.

Limit patent attorney involvement to what they're good at: making sure the claims are solid.

Make no mistake, however – I'll foreshadow that I recommend the hybrid approach. I'll show you how to prepare every element of your application such that you technically could go it on your own, but, much as I don't like it, I must strongly advise you to have a patent attorney review and edit at least your patent claims. That cost will be a small fraction of having a professional draft the whole application. You'll understand when we will get into the details later in Chapter 9.

The nature and merits of the relationship between patent attorneys and commercial entities is quite different than with independent inventors.

A higher level of involvement by patent professionals does make sense in a commercial setting. Commercial entities (businesses) customarily don't have sufficient work to employ dedicated staff to generate patents, and so they're stuck paying whatever it takes when they occasionally need to do so. Be that as it may, keep in mind that, because of the high cost, they very carefully evaluate ideas before they decide to shell out for a patent, and more often than not, they opt out.

1.7 TIMING

If your invention proves to be worth its salt, eventually you will reach a point where it will make financial sense to get a professional more involved, and here's where some judgment and timing comes in handy. To make something of your patent, you will either license it to someone else to use/manufacture or start your own enterprise, but in this process you have some latitude about exactly when you file for your patent. The patent process

> **If you play your cards right, you can use the time required to process your patent application to your advantage.**

does take a while, and generally a minimum of two years will pass between the filing date and the issue of a patent, and usually significantly longer, depending on how many iterations are required on the CLAIMS. You can expect it will be at least well over a year before you are required to respond to an "office action" (as correspondences from the USPTO regarding a patent application are known), so you can prepare marketing materials for licensing, a business plan, prototype, etc., and wait to file your patent until you are actually ready to start showing people your invention.

> **The day the USPTO receives your application establishes the patent's priority date.**

The date upon which the USPTO receives your (provisional or non-provisional) application is your "priority date", that is, the date from which you are officially credited with having established unique ownership of your invention. Effectively, even though the USPTO has not yet investigated the validity of your CLAIMS, you get an advance under the assumption that your application will ultimately yield a valid patent, and your invention is said to be "patent pending". Your initial filing does not really set anything in stone, and you can revise content in later submissions.

> **Filing your patent application sets the timeline of future related events.**

The point is, the longer you wait to file, the longer you can delay any potential need for professional help (that is, help beyond the basic pre-filing review of your patent to which I alluded above). For example, if your goal is to license your invention and you play your cards right, you can file for a patent shortly before licensing your invention, thereby leaving lots of time after the point where you now have some basis to believe your invention may actually be worth something to make decisions about increased involvement by professionals. From a minimum cost risk perspective, exclusive licensing is ideal, because once you sign an exclusive license, typically the patent process (including fees) is taken over by the licensee, who now has a vested interest in making sure the patent is solid, and better resources. (Since, in this case, all your eggs are in one basket, you must be very careful in selecting a licensee, however – more on this in Chapter 11.)

> **Delaying the filing of a patent application until right before you market your invention can leave most of the costs to the (exclusive) licensee.**

Your patent can, effectively, be edited (or more specifically, supplemented) after it has been issued, by filing a continuation (which is like filing a second patent with expansions on the same subject), but this gets more complicated. The first year following your patent application date or initial offer to sell (either your invention or a license-to-sell) is by far the most important period during which bringing on a professional can provide the most benefit, since this is the time during which you retain the most flexibility in regard to making changes and filing in other countries. At all times, your general strategy should be to keep your cost commensurate with the established value of your patent. Think in terms of risk vs. potential return on your investment.

> **Patents are easiest to modify during the application process, but they can also be modified after issue by filing continuations.**

Obviously, the longer you wait, the greater the chance that someone else will invent the same thing or something similar and file ahead of you. If your invention really is patentable, the odds of someone else filing a patent for the same thing any time soon should generally be small, since, by the definition of patentability, it must be non-obvious (more on this later). Nevertheless, there are seven billion other people in the world, and some of them are also smart (estimates regarding exactly what fraction vary quite a bit, but usually drop sharply around the time of elections). It really comes down to a judgment call that is specific to your invention and the field to which it pertains. If the field is young and it seems like there's a new patent popping up every time you check the database, clearly you should consider filing earlier. But once you file, the clock starts ticking, and you'll want to be in a position to move forward with your marketing strategy.

1.8 GETTING PRELIMINARY PROTECTION

As mentioned above, generally, your best strategy will be to file your patent application after performing the preliminary evaluation, patent searching, and prototyping tasks to be discussed in the following chapters of this guide. So if one waits until just before one is ready to pursue licensing agreements, how can the risk of having someone else come up with the same idea and filing a patent ahead of you be mitigated? Truthfully, if your invention is sufficiently obvious that this is a high risk, it is likely not all that much of an invention; however, there are clear exceptions. Naturally, for these cases, a provisional patent application is an option, provided you are thorough and follow through with it. Often, however, the above-mentioned tasks can be expected to take more than a year. So what then?

Provisional patent application protection often doesn't last long enough to span the development process.

Currently U.S. law grants priority to the first inventor of an invention, regardless of whether or not they are actually the first to file a patent application. That someone else should invent the same thing later, even independently, is technically irrelevant. (I say "technically" because having proof is one thing, having the money and legal horsepower to prove it in court is, unfortunately, quite another). Therefore, all that is essential is that you have some means to prove when you first conceived of your invention. In the professional world that means is known as an engineering or inventor's notebook – essentially a high-end doodle pad with fields for recording dates of entries and signatures of colleagues who act as witnesses to the invention. Through disciplined documentation according to a specific set of rules, such notebooks are admissible as evidence in a court of law.

Professionals in industries involving inventorship often maintain "inventor's notebooks" to document invention dates.

Naturally lawyers and many professionals in inventive fields are familiar with this process, and you can find considerable discussion on the internet on inventor and invention-related forums about how important it is for you to keep a methodical and up-to-date inventor's notebook. But, dear inventor, I expect there's an excellent chance that you've spent little of your home life scribing technical documentation, and, before reading this text, you've probably never heard of an inventor's notebook. So what do you do now? Firstly, starting one now won't help you much with something you've already invented. Much of an inventor's notebook's strength is derived from its chronological trail of notes, other invention concepts, etc. that firmly anchor the invention within a continuous body of work. An inventor's notebook with one invention in it that doesn't connect with a contract number or some other cross reference doesn't really provide much evidence about the invention's vintage.

Most independent inventors aren't doing enough inventing to make an inventor's notebook practical

Frankly, the idea that regular folk, even the kind that invent things, would keep, maintain, and periodically have witnessed an inventor's notebook is pretty silly (and you should roll your eyes when you see internet puff-ups talking this up as something that would be expected of you). Another common, and often simpler (and actually stronger) means of documentation in common use in industry is what's known as a "Memorandum of Invention". Simply write yourself a memo (preferably using "Memorandum of Invention" in the title) that describes the invention. Illustrations should be included, along with the name of the inventors and the place where the invention took place. Since you probably don't have an ISO 9001-approved document control system at home, you may authenticate the date of composition of the memo simply by having it notarized. Most banks will provide this service for less than $10, and often for free to account holders. Remember, therefore, to include signature lines for yourself and a witness. (Most notaries will not act as a witness, by the way – you must bring one with you).

A notarized "Memorandum of Invention" is often more practical than keeping an inventor's notebook for the independent inventor.

In Section 2.2 you will learn about recent legislation which essentially makes inventor's notebooks obsolete as of March 15, 2013, when a law will go into effect granting priority to whatever inventor files the first patent application.

1.9 PROMOTING YOUR INVENTION

There are numerous people and organizations out there who will offer to help you promote and patent your idea for a fee. If you check out the USPTO's information section, you will see warnings to approach such organizations with caution. I'll tell you not to approach them at all. The USPTO will warn you that many of these organizations are shams. I'll tell you that they all are. Keep in mind that the USPTO needs to be cautious with its words in regard to scammers, because charlatans are just as happy gouging the government with frivolous lawsuits as they are fleecing you, and their pockets are deeper. I've already caveated everything in this document as opinion, so I'll call it as I see it.

Legitimate professional invention promoters are fictional.

I've never seen one of these promoters that didn't turn out to be a scam. Friends of mine who have used them have always been sorry, and the ones that last long enough usually end up prosecuted for fraud. I think that some of them actually may start with good intentions (don't get me wrong – most really are rotten through-and-through), but the fact is that even if they start out genuinely hoping to help inventors make something out of their ideas, they very quickly find out that there really isn't all that much they can do. You pay them to promote, so they promote, but their efforts really won't substantially increase your chances, after all who are they that you are not? So they start puffing up their credentials, their claims of access to industry and past successes, and pretty soon what they can provide doesn't match what they claim they can provide. But I'm being nice. In truth – most of them just simply are shameless villains.

Invention promoters come in two breeds – those that first practiced to deceive, and those that just inevitably finished in the same place.

Expect to make a prototype.

Effective promotion of an invention can usually be accomplished for little more than the cost of travel.

Anyone who says different is usually trying to sell you something.

In many cases, the most significant step you can take to improve your odds is to create a prototype. There's nothing more convincing than a tangible example, either to potential licensees or to investors/potential partners (if you're planning on starting your own business around your patent product). If you do plan to launch a company to take your invention to the market on your own, all I can say is good luck. Licensing an invention and running a startup are two completely separate endeavors, and I have little experience or advice on the latter, but my underlying philosophy remains the same. Sacrifice time and effort in whatever measure you can spare them, but never the financial security or wellbeing of your family. Coming back on topic, the rule does not change for prototypes. Expect to make a prototype, and adopt a strategy where you do as much of the work yourself as possible. We'll come back to prototyping when we get to Chapter 4; so for the moment we'll just leave it at that.

Basically, if promoting your patent costs you much more than travel expenses, you're probably on the wrong track. The young typically don't have that much money to spend; the less young have better things to spend it on. Hollywood will always portray individuals who risk everything for their dream as heroes (gag), but at the end of the day, fictional characters don't have real families to feed, and their successes don't need to be counterbalanced by a statistical sea of less-than-successes. To risk destitution on an invention is just plain irresponsible – unfortunately, irresponsible people are good shoppers, so they'll always be fashionable in Hollywood.

To Patent or Not to Patent?

2.1 SHOULD I PATENT MY IDEA?

Few people ever patent an invention.

...but perhaps more than you'd expect.

...and certainly more than should, judging from the numerous silly things you'll find in the patent literature.

Per U.S. patent law, any person who "invents or discovers any new and useful process, machine, manufacture, or composition of matter, or any new and useful improvement thereof, may obtain a patent." But should you? The truth is that there are far too many wanna-be's filing worthless patents. Don't get me wrong – I'd take one of them any day over ten do-nothings who whittle away their lives on their sofa, remote in hand. The fact that you have an idea and you're now actually doing something about it puts you in a pretty elite crowd, and I like you already. Let's put it this way – about 7.2 million patents have been filed since the first U.S. patent was issued in 1790 (signed by George Washington himself). The current U.S. population is about 300 million, and has doubled about once every thirty years; so in the time the U.S. has been issuing patents, about 600 million people have lived. (Yes, there are actually as many citizens currently alive in the United States as have ever previously lived, but that's another (rather scary) topic.) That equates to an average of one patent being filed for about every eighty people, but, accounting for the fact that inventors tend to contribute multiple patents in a lifetime (the Idahoans lead the nation when last I checked, by the way), certainly not more than one in three hundred people ever file for a patent.

We will now commence with the first of several steps in the process of moving toward filing a patent application. In this chapter, we will simply begin by making a basic assessment of the quality and worth of your invention. We will not need to go overboard on this, but your first expenditure of time and effort related to your invention really needs to be to seek a modicum of perspective.

2.2 CAN MY INVENTION BE PATENTED?

To be patentable, an invention must be novel, useful, and non-obvious

Not all ideas that are patentable are worth patenting, but let's start by going over what is patentable. When we typically think of an invention, we think of a physical device or machine, but practically any new idea can be classified as a new or improved "useful process, machine, manufacture, or composition of matter". Clear exceptions are laws of nature, physical phenomena, and abstract ideas. Basically, if you create it, you can patent it; if you discover it, you can't. Beyond these criteria, there are some key rules – the USPTO grants patents only to inventions that are (1) novel, (2) useful, and (3) non-obvious.

2.2.1 Novelty

Clearly your idea is new to you, but to be patentable, it has to be new to everyone. It is not sufficient that no one has previously patented your idea. The law states that you may not receive a patent if:

One may not patent an invention simply because it has not previously been patented.

"(a) The invention was known or used by others in this country, or patented or described in a printed publication in this or a foreign country, before the invention thereof by the applicant for patent,"

or

If anyone has sold or published anything about it before, it does not meet the novelty requirement.

"(b) The invention was patented or described in a printed publication in this or a foreign country or in public use or on sale in this country more than one year prior to the application for patent in the United States."

If you are awarded a patent and someone else later shows they published proof of the same thing first, your patent is null and void.

If it can be shown that anywhere else in the world someone already documented or sold your invention (or something similar with only obvious differences, see 2.2.3 below), you cannot get a patent. If you did get a patent, and what is generally referred to as "prior art" subsequently turns out to demonstrably to have existed prior to the filing of your application, your patent will be invalid. The second clause above (b) refers to an additional condition that states that even if you are the original inventor, if you or anyone else patents your invention elsewhere in the world, writes about it in any published forum, or offers to sell it, from that point forward you have exactly one year to file your patent application. If your invention is disqualified by either of these criteria, you're out of luck.

On March 15, 2013 the U.S. will move from being a "first-to-invent" to "first-to-file" nation.

If the invention is novel, traditionally, U.S. law has given priority (basically ownership of the intellectual property) to the first person to conceive an invention, regardless who files first for a patent application, which you may correctly suspect has occasionally left much to be desired in terms of concrete proof. That is scheduled to change with the signing of the Leahy-Smith America Invents Act into law on September 16, 2011, however, whereby eighteen months thereafter the U.S. will officially become a first-to-file nation; that is, whoever files the first application will have priority. Now, if you search in the right places, you will encounter many words applauding or lambasting the new legislation, and there are certainly items in the bill that will prove ill-conceived or feebly implemented (this is Congress we're talking about after all); but, in truth, from a practical standpoint the change from first-to-invent to first-to-file will be of little consequence most of the time.

Even still, only actual inventors can receive a patent.

The new law grants the right-to-exclude to the first legitimate applicant for an invention – the pivotal word being "legitimate". The new law does not enable someone who learns of your invention to run to the USPTO and file a patent application ahead of you, for stealing someone else's idea only makes one a thief; and the USPTO can grant patents only to actual inventors. The only circumstance where a bona fide inventor stands to lose is if someone else independently and contemporaneously happens to invent essentially the same thing, in which case whoever files first will win out; but someone must, and that seems as fair as anything else. Importantly, the new stance does protect you against a potential circumstance where, ten years after you patent your invention and make it big, someone shows up with an old scrap of credible documentation demonstrating they actually developed the same invention a year earlier than you did but never did anything with it, and now claims damages and takes control of your patent. Another benefit is that U.S. patent law is now better aligned with the large majority of the other nations of the world, which paves the way for greater synergies, and perhaps even someday real reciprocity, in international intellectual property protection.

2.2.2 Usefulness

The USPTO has largely abrogated its role in preventing useless patents, but to be useful, an invention must work.

Clearly one must draw wide margins when passing judgment as to what is and is not useful, and, understandably the USPTO has largely abrogated its role in preventing the patenting of useless inventions. These days, they'll just smile, take your fee, and issue you the requested useless patent. One key feature of this requirement, however, is that it excludes inventions that don't really work. If I try to patent a device that claims to grant immortality by strapping a TV remote to my forehead or allows travel through the great expanses of space using warp drive, I won't receive a patent unless I can convince the USPTO reviewer that my device/method works (or more specifically, I need to at least persuade the reviewer that he/she is not confident that it won't work). This couples with a requirement that we will address in Chapter 8 for the patenter to provide at least one example or "preferred embodiment" of how the invention may be reduced to practice. If you can't show at least one way you would actually make your invention, then you haven't invented anything. So before you overnight an application for a hyperdrive system showing how a spacecraft can do superluminal surfing on warp waves to the USPTO, make sure you know and can demonstrate how to generate warp waves.

Now, do be prepared to stand by your guns. The USPTO employs qualified staff (though it cannot help but be somewhat of a mixed bag), and your reviewer will have expertise relevant to the field of your invention, but no one knows everything. I have had the occasion where a patent reviewer initially objected to my patent on the basis that it would not work as illustrated by the preferred embodiment, when, in fact, I had already been tinkering with a working prototype for a number of months. I expect this is rare, but if this does happen, just patiently explain in your response to the office action. After all, you've spent a great deal of time thinking about your invention, but to the reviewer it's entirely new (hopefully), and he/she is being hit with it cold. (Here's one place a working prototype is very useful.) You will probably end up being required to send a signed affidavit to the effect that you have, indeed, successfully reduced your invention to practice, as I did. Don't you dare lie – I haven't ever tried committing fraud, but I expect if you do, bad things happen in the end. When you go to federal prison, there's no early release for good behavior. These days, you won't do all that much time, but you'll serve the sentence delivered.

> **It has been known to happen that the patent reviewer mistakenly thinks your invention can't work.**
>
> **There's nothing like a working prototype to squelch an objection such as that.**

2.2.3 Non-obviousness

Your invention must be non-obvious; that is, it must not be obvious to those "skilled in the art", i.e. people with sufficient experience and/or education in the field of invention. If your doctor wouldn't have a clue how to make your new waterless toilet, who cares; it's what your plumber would think that counts. That said, whether or not something is obvious is even grayer than whether or not it's useful, and the standard of non-obviousness has evolved to be increasingly liberal over time. Generally speaking, a patent answers the question "how do I do X?" or "how could I make X?" If "X" is something someone might probably not know how to do or make without my telling them, it's not obvious.

But now let's consider the question relevant to a patent, say, on the fabrication of a lawn-mower using an electric motor where, just hypothetically, all known art consists of those employing gasoline engines. If one considers the question to be "how do I make a lawnmower powered by an electric motor?" it sounds pretty obvious that I just replace the gasoline engine with an electric motor (and I'm inclined to think this one does count as obvious). But what if the question I asked instead was "how do I make a quieter lawn mower?" Now clearly different people would answer this question differently, and only some of them might conclude that the best approach was an electric motor, but others would not. So it is quite plain that what counts as obvious is pretty subjective, and these days the USPTO is not very restrictive in this regard. As long as you're not trying to patent something like a change in color or size, if it meets the novelty and usefulness criteria above, you can probably take non-obviousness for granted. As we observed in Section 1.3, it's up to you to enforce your patent, so generally the attitude these days is if your patent really is lame, you'll get KO'd in court if you ever try to make an issue of it.

> **Unfortunately, the USPTO has also evolved to give you great latitude (and then some) on the issue of non-obviousness.**
>
> **Whether or not something is obvious can be very much a matter of perspective.**

> **Oh, one more thing:** Apparently 42 U.S.C. 2181 (a) of the Atomic Energy Act of 1954 excludes the patenting of inventions useful solely in the utilization of special nuclear material or atomic energy in a weapon, so no nukes. Fair enough?

2.3 IS MY INVENTION LAME?

Mastering the ability to be objective with your own ideas will be a great life-long asset.

Not to be rude, but I don't know how else to pose the question. Difficult though it may be, you need to try to be objective with your own ideas. That means you have to be able to ask yourself the hard questions, and answer them honestly. Basically, what I'm suggesting is that you be your own harshest critic (which is way better than letting someone else have the job). Ask yourself the following questions:

1 Would I use my invention? If you won't, then no one else will either.

2 Would people laugh at me if they saw me using my invention? Hey, ultimately the market is going to decide if your invention is any good or not, and it's hard to market something that looks plain silly. Of course, there was Henry Ford's horseless carriage, but, really, if your invention looks ridiculous, then chances are, it needs work. It may not be a total bust, but you should devise a way to make it cool.

3 Can I easily think of a better alternative to my invention? If you can, then start over with that idea. You really need to put some time into what I call refining your invention. If your invention still looks just like that original napkin sketch, be worried. Usually, after the initial light bulb suddenly illuminates your brain and you shout "Eureka!", improvements come to you initially quickly, and subsequently at an asymptotically diminishing rate. Once you've thought about your invention for at least ten hours or so without coming up with any new improvements, it may be ready. That's ten hours of actual concentration, not 10 hours from the time you came up with the concept, and I'm talking here about a relatively simple invention. (On one of my inventions, I eventually came up with a critical-to-success subcomponent innovation two years (and I can't tell you how many hours of pondering) after the initial invention).

4 What do my trusted friends and family think of my invention? I touched on this in Section 1.6, but I should reiterate here that you cannot rely solely on your own judgment. I've said that you should be your own harshest critic. A wife, brother, or sister is usually good for a fairly brutal review too. Regardless of who it is, you need to bounce your invention off of closely trusted family and friends – and when I say closely trusted, I don't mean that they need to be trustworthy so as not to steal your invention. I mean you need to be able to trust that they won't spare your feelings.

I have received the occasional suggestion that this chapter should be more expansive; that focus groups and surveys should be discussed, etc. Of course it's only natural to wish for a way to know an invention will be a market success before going to all the effort required to get a patent, but the unfortunate truth of the matter is that extensive market research is a very large undertaking in-and-of-itself, and hardly represents a means of avoiding unnecessary expenditure of resources. Further, any psychologist will tell you that where and how a question is asked greatly affects the response; and to design a meaningful survey requires considerable skill and effort. Even then the data will only be as good as your ability to interpret it (for responses are composited with many extraneous influences)...and no matter what anyone says, there are no sure things.

Worthwhile market research can be a much more formidable project than filing for a patent.

So are market surveys worthless? Of course not, which is why large firms often spend millions on them before launching a product to the marketplace (the cost of which makes both patent and market research costs look like chump change). But good ones are expensive and follow, not precede, intellectual property protection (since they universally involve disclosure of the product outside the company). So the question isn't why invest time, energy, and money in a patent before doing extensive market research, but rather why invest an even larger amount of time, energy, and money on market research of an unprotected idea, especially if your plan is to license, in which case the licensee will save you the trouble. In the same vein as Section 1.9, do not give your money to someone offering to research the marketability of your idea unless you just like giving money away. To gauge whether your invention is truly worthy of pursuing, you simply must open yourself to the possibility that it isn't. You needn't stray far from home to find those who will gladly let the wind out of your sails; and if you know them well, you may even understand what it is they are trying to tell you.

Market surveys, if done, generally come after intellectual property protection, not before.

2.4 Is My Invention Worth Patenting?

The next question before the jury is "can my invention be profitable?", not "will it be?". Is it even possible? While one cannot know for a fact that an invention will have financial success, some quick math may tell you it won't. Case in point – on one occasion in my career as a propulsion engineer (call us rocket scientists if you mean to flatter), I came up with a nifty little device that provided a simple means of mixture ratio control for a bipropellant rocket engine. We needed to be able to talk about it, so we quickly filed a provisional patent (recall Section 1.4). As always, before we knew it, a year was coming up and we needed to make a decision as to whether or not to go to the expense of pursuing the full patent. Now this was a fairly simple device (that was the point) that could save on the order of $10k per subsystem – Wow! Not really, you see, each of these subsystems comes at a typical price tag of several million dollars, so $10k is chump change. Even if every launch carried the component (but it wouldn't actually be applicable to all systems), at thirty to fifty launches annually world-wide we're talking $300k to $500k per year, for a savings on the order of five one-thousands of a percent on a typical multi-$100M mission budget.

Before even undertaking the (larger) task of determining if your invention is original, first consider its potential worth.

Consequently, you can see that once we took a look at the big picture, we dropped it quick.

You really owe it to yourself to find out how much you really can hope to make from your invention before you do anything else. The potential payoff needs to be sufficient to provide a reasonable reward for both you and any partner(s) that would be involved, plus a licensee if you intend to include one in your plans. Handily, these days that data is typically no further away than the internet. Research the relevant industry – find out how many real potential consumers there are for your invention by looking at products that currently fill a similar role, what the total sales volume is for such related products, the relevant industry annual gross sales, etc.

> **Learn about the total annual revenue of similar products – usually the internet is the only source you'll need.**

Another thing that you will need to think about is how much profit you expect your invention will yield on a per unit basis. We're talking net here; that is, retail minus the production cost. Don't worry if you don't have a clue how to gauge this – most readers won't, but fortunately we only need a rough order of magnitude estimate. First, you need to make a reasonable judgment about what retail price the market will bear for your product. Find something similar, say the product you hope to replace, a product that plays a similar role in an alternate field, whatever. Just find something with which you can compare – we're doing quick math here. Don't worry about trying to be accurate to several digits, you're really just trying to figure out if you think your product can net $100k/year, $1M/year, $100M/year, (or maybe negative $100M/year?); you know, just an order of magnitude. Try to pick something that you believe also targets a market of similar size. You can plus up a little if you think people would pay more for the improved performance of your invention, but don't overdo it. Remember, the higher the price, the lower the sales volume, so I generally recommend just using the price of the immediately most comparable product.

> **Use existing products to gauge what retail price the market will bear for your invention.**

For production costs, again pick something doing similar sales volume, but now you're interested in something of comparable complexity and method of manufacture to your invention, but not necessarily fulfilling the same need. Obviously it's easiest if this is the same as the product you compared to for your retail estimate, but if your product is intended to replace something much simpler to make, you'll probably want to compare with something else. Even if you pick an alternate product, still pick mechanically similar things – compare electronics with electronics, pneumatics with pneumatics, cosmetic with cosmetics, etc. Lacking anything better (which will be the case for most readers), you can just take the retail price of that item, and multiply by one third. This is your expected production cost/unit. Subtracting your expected cost from projected retail price, you now have a very rough estimate of net yield per unit. Keep in mind, you'll only take home part of that. Another third will go to marketing and distribution.

> **Production costs may quickly be estimated by comparing to a product of comparable complexity and method of manufacture.**

If you license, expect personally to make less than 20% of the retail price (and most typically less than 10%). That's a big range, and I wish I could give you a formula that would pin it down specifically for your product. Obviously if you walk into a board room, give a pitch, and receive an offer, you'll want to know whether or not it's a good one. The truth is, that's very hard data to come by, and even if you knew what was typical there's still case-specific development risk to be considered and sometimes a by-the-seat-

Licensing will typically reduce your yield to <10% of the retail price.

of-your pants judgment call to be made. The greater your royalty, the less incentive the licensee has to promote your product over other options (and they won't have all their eggs in one basket if they can at all avoid it, even if you do). We'll talk more about this in Chapter 11, where our focus will turn to how to pursue a license agreement. For now, the point isn't to spend gobs of time doing exhaustive research – keep it simple. But you would be surprised how many inventors have come to me with an idea they are very excited about patenting having given no thought at all to profitability. Not even all really good inventions have profit potential. Some wonderful things have been brought into the world that completely lack commercial value, in which case no patent is required.

Action Step 1

Stop here and search the internet for data relevant to your invention's profitability. Record your results in a table like the one shown in Fig. 2-1. A blank table is provided in Section 1 of the Application Workbook.

Make sure you filled in the "Basis of estimate boxes at the right. In particular, you want to just leave yourself enough of a note so that if you want to, you'll be able to remember where your numbers came from and why you thought they were reasonable. Since we're just looking for an approximate magnitude, you may use 5% of the estimated retail price for net return per unit.

Total units/yr I can expect to sell: 60M	Basis of Estimate: One unit per consumer. Market saturation – everyone but me, phased in over next five years U.S. population = 300M (per http://www.census.gov/)
My expected net return per unit is: $15	Basis of Estimate: Complexity similar to standard cell phone.
My estimated net annual return will be: $900M/yr	

Fig. 2-1 Estimated annual return and BOE on Portable Psychotronic Mind Control Implant

If you have not already done so, you should download the *Patent Application Workbook* now. Here's the link:

http://www.inventionpatentinformation.com/freepatentworkbook

Now that you've finished are you still interested? If your answer is yes, move on to this short quiz (you can click the boxes in the Application Workbook to check them off).

Action Step 2

Check all that apply:

- [ ] My invention is novel.
- [ ] My invention is useful.
- [ ] My invention is non-obvious.
- [ ] I have discussed my invention with at least two trusted confidants who agree that it is not lame.
- [ ] My estimated return is worth the effort.

You need a perfect score or odds are that patenting your invention will be a waste of your time and money. Did your idea pass the test?

YES!

Excellent to hear it! But let's not break out the champagne just yet. So far all we've established is that your idea merits going on to Chapter 3, which represents an escalation in your time commitment.

If you're perceptive, when you came to the question of novelty in the questionnaire above, since you answered yes, you're here, but, hopefully you're fully aware that the answer was actually, yes, I think so...but you don't really know for sure. What you probably meant at this stage is that you are not aware of anywhere your invention already exists, but you really can't say that you're confident that it, in fact, does not already exist. There's a big difference between not knowing that it does, and knowing that it doesn't. We'll deal with that in the next chapter.

NO.

Well, believe me, you're not the only one. I wish I could say that every idea I've had was a winner, but sometimes they just don't make the cut. The good news is that you probably just saved yourself a ton of wasted time and at least $1-2k.

DO NOT BE DISCOURAGED.

If you invented one thing, you can bet your next big idea will be along shortly. The inventive spark doesn't touch everyone, but usually where it does strike, it strikes again. Unfortunately, it's difficult to invent things while sitting on your sofa (except sofa products) – complacency is creativity's greatest enemy. Chances are your idea came to you because you were doing something and started thinking about how it could be done better. Don't lose that. Be a doer, and if you truly are an inventor, you won't be able to help yourself.

3 Conducting a Patent Search

3.1 Let's Find Out If You're the First

More often than not, when you have an idea for an invention, you'll find it's already taken.

You are still not ready to bet the family farm. (If you read and understood Chapter 2, you should have immediately reacted to that statement – betting the family farm will never be an option). Hopefully, in Chapter 2 you took a hard look at whether your invention really is inventive, whether or not people would buy it, and whether or not you'd make anything if they did. Assuming you did, maybe you're really on to something. The trouble is, if you're onto it, unfortunately there's a pretty good chance someone else is onto it as well. If you do end up finding out someone else has indeed already patented your invention, try not to feel bitter. If you met, you'd probably like him or her. After all, more than anyone else, it is most likely that he/she thinks your idea is really swell.

3.2 Patent Search

In the quiz at the end of Chapter 2 you indicated that, as far as you know, your invention is novel. Well, it's time to do your best to find out for sure. You just can't go spending gobs of time developing your invention without first being absolutely certain that it isn't already out there – and that's

Don't get attached to your invention until you have verified that it's novel by completing a thorough patent search.

It takes much longer to verify the absence of prior art than to find it.

The USPTO provides a comprehensive online database of every existing U.S. patent you need to worry about.

You will have to install a special browser to view patent image files, but freeware is available.

where the patent search comes in. Once again, you have the option to engage a professional to do a patent search, and, once again, I'm going to tell you that the cost way outweighs your current level of confidence in the profit potential of your idea. It would be one thing if the service offered was to verify that your invention was novel, and you only had to pay if no prior art turned up. But, you pay either way. Also, in order for someone else to do a patent search for you, you'd have to explain to them your invention, which means you'd have to jump in and get a pretty good start on a written description of it – but as I indicated before, you will want to avoid investing the time to do so until you know it's worth the effort. Remember, up front your focus should be exclusively to evaluate your idea's worth and patentability. Once that's out of the way, you can move forward without the dread of being derailed after many hours of wasted effort.

When you pay someone to do a patent search, what do you suppose they do? Nowadays, they hop on the internet, pull up the USPTO website (or another site offering access to patent databases), and do a keyword search. If you can Google or Yahoo, you can do a patent search. Chances are, the whole search won't actually take that long. Unfortunately, that's because you may find what you're hoping not to find very quickly. As the hours drag on (and how long it takes will vary quite a bit, depending on how old and popular the field of your invention), take consolation in the fact that as the time goes by without your finding your invention already patented, the odds of it actually turning out to be novel are steadily improving.

There's this guy who sells compilations of patents for different fields on CD on Ebay. Now, his customers do get exactly what they pay for...but, the thing is, the USPTO has already provided you a much more effective tool, absolutely FREE! (Did I mention I'm a fan.) The USPTO search engine is powerful, and you can distill it down to the same list that guy sells on Ebay with one keyword search in less than a minute (where do you think his lists come from?). Tell me you didn't give that guy your money. If you did, you're about to feel pretty silly.

To view patent image files, you will need to download a plug-in for your browser. The USPTO stores patent image pages as 300 dpi TIFF's to conform to an internationally agreed upon standard, but these are not just regular TIFF files. Lossless CCITT Group 4 compression is used to keep the database size down to a trim four terabytes, but this means that your garden variety image display software won't work, and, in fact few things will. The USPTO recommends either one of two plug-ins called InterneTIFF™ and AlternaTIFF™. The freeware version of InterneTIFF™ expires just three days after installation, after which the functionality becomes very limited. AlternaTIFF™ doesn't have that annoying feature, so I recommend it, but you do have to register to enable it after downloading (but that's free, except maybe you end up on a junk mail list). Other support is offered for Macintosh and Linux, which I have no personal experience with; so if one of those is your flavor, all I can do is wish you luck.

> **Action Step 3**
>
> Download the plug-in of your choice now. You will need it in just a few moments. Here's a link to the USPTO's plug-in links page:
>
> http://www.uspto.gov/patft/help/images.htm

Freepatentsonline is becoming the free search tool of choice because of its access to a broader set of databases.

Other online resources, such as Freepatentsonline.com (http://www.freepatentsonline.com), also provide access to keyword searchable patent databases, and all have there pros and cons. Freepatentsonline, for instance, requires no special plug-in other than Adobe Acrobat, but isn't as fast as the USPTO's site. Freepatentsonline does provide the handy ability to simultaneously search the U.S. published pending applications database along with the U.S. patent database, which can be a time-saver, and can also pull up foreign patents. More recently I have been favoring Freepatentsonline because of it's broader access. As the general approach to conducting an effective and strategic (with respect to use of time) search is independent of the choice of search tool, we'll stick with the USPTO's site for general illustration as we move forward in this chapter.

3.3 PART I – KEYWORD SEARCH

Start your search with keywords that naturally describe, and are very specific to, your invention.

We'll go on over to the USPTO site in a moment, but first let's talk about a time-saving strategy. For what you're doing, the keyword search is going to be most useful, and generally you can just go ahead and search all available fields. What you're first going to search for are keywords that are not very general, but the most specific words that you would expect someone to use when talking about your or a similar invention. If your idea is already in the art, we want to waste as little time as possible finding it.

Your keyword searches should progress from specific to general.

Of course, if the obvious keywords do not immediately pull up an existing patent for your idea, you're not done. Next you will need to move on to more general words. Sometimes patents intentionally use somewhat off-center wording to make them harder to find. The more patents you search, the more certain you will be that you won't be disappointed a year after you file your application by a dismissal of your application on grounds of non-novelty. Since you're moving from smaller, more specific searches to ones that are larger and more general, take notice of your browser's re-coloring of links you've already visited to save you the trouble of reviewing some patents twice.

You have a lot of ground to cover – you can expect to review hundreds to thousands of patents (but usually closer to hundreds, unless your invention relates to toilets – it seems like everyone has at least one toilet invention). Once you've pulled up a list based on a keyword search, just start at the top and start going down. Some of them will be obviously different than your invention just by the title, but most won't. In general, when in doubt, click it. Fortunately, here's a handy trick (which most people would figure out pretty quickly on their own) so you won't have to read almost any of them. Napoleon Bonaparte may have made a bad turn somewhere along the line, but not when he said "a picture is worth a thousand words." When you open a patent, click straight through to the images, and then skip to the drawings. Scanning the drawings, you will find that you can dismiss 99% in less than a minute without having to read a single word. So, you have hundreds to thousands of patents to review, but 99% of them will only take you about a minute.

> **Usually the fastest way to dismiss irrelevant patents is by first glancing at the drawings.**

Make a list of all patents that are not totally irrelevant, and put down a one-line note about what is similar to your invention Generally, your goal will be to make it easy for you to go back to these references when you compile the "CLAIMS" section of your patent application. Also, it will save you time later to keep in this list patents that you intend to reference when you write the "BACKGROUND OF THE INVENTION" section of your patent specification. Keep track of anything that illustrates limitations that your patent overcomes or alternate attempts to alleviate the same problem. You won't necessarily end up making reference to all or even any of these in your patent application (there is no requirement for you to site any prior patents), but if you do come to a point where it would help you to use a reference, you won't want to have to go back and look for it from scratch. You will understand more about why this will come in handy when we come to Chapter 5.

> **Make notes about prior patents that attempt to solve the same problem as your invention or otherwise may be useful to you later.**

If you are using a tool that does not allow you to do so simultaneously (e.g. the USPTO's patent search engine), remember for each search phrase you should search both the published application and patent databases. As you go, you should also keep a log record of all search expressions you use. This will save you from accidentally doing redundant searches, and will help you brainstorm additional search parameters.

> **Keeping a log of expressions you have searched will help you be thorough.**

Now let's click over to the USPTO's site and do a quick example search. Here's the link:

http://www.uspto.gov/

On the left hand list, under "Patents", find the link for patent searches and click through. This will take you to a page with a list of search types. The most powerful of these is the Boolean search engine under "Advanced Search". I encourage you to explore this, and you will probably find that a quick glance at the several examples shown will immediately educate you on how to use it. Most of the time, you won't actually end up using this, because the "Quick Search" option allows you to do simple Boolean two-phrase searches, which will probably be all you need. For an example search, let's go back to our notional "Electric Powered Lawn Mower" invention. On the quick search page, start by typing in the words "lawn mower". At the time of this writing, this search brought up exactly 3363 entries. Now, at our estimated average of one minute per element, you could just browse this list, title by title – It'd take you about 56 hrs. That's a lot of work, and, hey, you probably already work at least 40 hrs/week. However, most of these will have nothing to do with an electric lawnmower, so let's narrow the search by clicking back, adding the word "electric" into the second keyword box, and leaving the Boolean option on "AND". That narrows the field to 997 entries. That would still take you about 16 hrs to search, but that's getting more tractable.

> **Use Boolean keyword combinations to first peruse high-match-probability subsets of the total relevant prior art.**

Did you notice that the box next to the "Refine Search" button just added the word "AND" (in caps) and our second keyword? Try this: Let's say we're not just interested in any old electric lawn mower, but a battery-powered electric lawn mower. In the box next to the "Refine Search" button, type the words "AND battery" and click the button. Now the field is down to 299 patents, and that's a reasonable starting list for a first pass. Note, I said first pass – If I really were patenting a battery-powered lawn mower, I'd want to take a glance at all those electric-powered lawn mower patents, because clearly I may find some limitations to what I might be able to claim there. But first, I'd try a number of very narrow searches, such as "noise AND 'lawn mower'" (because I might be interested to see what electric lawn mowers are filed as noise-reducing lawn mowers), "motor AND lawn mower", "motor AND lawn AND trimmer" – you get the gist. You also need to account for alternate spellings, such as "lawnmower" and lawn-mower". Parentheses and Booleans can save you time, e.g. "("lawn mower" OR lawnmower) AND battery".

> **The USPTO's powerful search engine includes a refine search tool which will often save you the effort of drafting long Booleans.**

Try to avoid keyword combinations that are arbitrarily narrow. Above, we actually made just such an error when we searched for "'lawn mower' AND electric AND battery." Clearly here I'd be interested in any patent containing the words "lawnmower AND battery", so adding the word "electric" just eliminated patents that I actually would want to come up in the list – 121 in the U.S. prior art to be exact.

> **Avoid accidentally over-constraining your searches.**

Take clues from the entries that come up on one search to help you think of searches you probably want to also conduct (jot them down when you see them, or you'll forget). In our last search, right at the top I noticed someone entitled their invention "sod cutting device". I'd just call it a lawn mower, but "'sod cutting' and battery" brings up another four patents.

> **Search results can often help you pick up other search words or phrases you may not have thought of on your own.**

Before you get started make sure to set your browser's visited link history retention span to a sufficient duration such that it won't start forgetting which patents you've visited before you've finished (I'd say at least twenty

Set your browser to remember which links you've visited for at least long enough to complete your patent search.

If the most specific keywords you can think of to describe your invention bring up many thousands of links, your idea may simply be too generic.

days, but if your time is limited make it longer. It won't really matter how long you set it for once you're done and set it back to normal). As I said before, since you'll be conducting multiple searches, it's sure nice if the links to the ones you've already viewed stay marked by a different color until you complete your entire search, so you won't end up looking at some patents twice. What did we ever do before the internet!? (I try not to think about it.)

In our example, we searched for a pretty generic invention, so a lot of stuff came up. Hopefully, your idea is quite a bit more unique than our "electric lawnmower", and your searches won't pull up quite so many hits. If your idea does bring up thousands upon thousands of past patents, be warned that it really may be simply too generic.

3.4 PART II – US CLASSIFICATION CODE SEARCH

Patents filed prior to 1976 can't be accessed by keyword searches.

Now there is one catch. Did you happen to notice that message on the search page where it says that patents filed prior to 1976 can only be pulled up by "issue date, patent number, and current US classification"? Only patents filed in 1976 or later are searchable by keywords. A little over half (3,930,270) of existing patents, however, were filed prior to this. They are available for you to view, but they only exist as image files (vs. both as image and text files), so the keyword search engine can't look at them. Nevertheless, you may need to, and here's how: Unless you're a powerful psychic, trying to guess the issue date or number of older patents relating to your field of invention is going to be kind of slow. Fortunately, the "US classification" is a searchable field for any patent ever filed in the United States and basically indexes the subject matter of the invention by a code, comprised of a "class" and "subclass" as illustrated below:

The only practical way for you to search the pre-1976 art is by US Classification code.

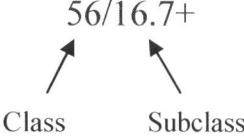

The USPTO provides an index of US classification codes on its website.

You mean you don't know what the US classification is for your invention? Here's a link to the USPTO's class index:

http://www.uspto.gov/go/classification/uspcindex/indextouspc.htm

The example code above happens to be the one most immediately relevant to motorized lawn mowers, but there are a number of subclasses that one would also want to have a look at. If you follow the link above, click to the "L's", and scroll down you should find a list that looks like this:

Lawn Aerator	172 / 21+
Grass PLT / 388+Mower (see harvester)	56 / 229+
Bar handle hardware	16 / 437
Circumferentially spaced blade	172 / 552
Design	D15 / 14+
Disc type	56 / 255+
Motor operated	56 / 16.7+
Mower-type tongue & crossbar	16 / DIG 38
Reel type	56 / 249+
Sharpener	76 / 82.1
Sharpener abrading in situ	451 / 419+
Small engine starters	123 / 179.26
Snowplow combined	37 / 243
Tool driven, laterally extended	172 / 121
Bar or blade	172 / 121

So, clearly for my electric lawn mower I'm interested in searching the entire patent database for the "Motor-operated" US classification code, but depending on the details of my invention, I will probably also want to check out some of the others.

It will generally be quicker and more effective to collect US classification codes from the references that come up in your keyword searches.

A more precise and convenient way to identify US classification codes for your Part II US Classification Code Search will be to make note of them during your Part I Keyword Search. Naturally, all of the more recent patents will also be classified by keywords. When you find and list patents that you consider relevant to your own, you should also record all of their classification codes. This way, once you have completed your Part I search, you will already have a list of classification codes to check out when you go on to Part II. (You should still browse the classification code index, to make sure you didn't miss anything.)

Use the USPTO's Advanced Search tool to find prior art by US classification codes.

Once you have collected US classification codes, to do searches, instead of using the "Quick Search" option, click over to the "Advanced Search" and type "CCL/XXX/xxx", where "XXX" is the classification code and "xxx" the subclassification. If you want to save yourself some time, you can bundle together a list of the codes you're interested in, and search them all at once by typing:

CCL/XXX/xxx OR CCL/YYY/yyy OR CCL/ZZZ/zzz…etc.

If you're certain that it's impossible for your invention to have been preempted before 1976, you can skip the US classification code search.

Of course, if your invention relates to a field where you can be absolutely certain that no prior art exists before 1976, you can skip the Part II search.

For example, if you've invented something for use with a PC that wouldn't have applied to a typewriter, you're good to go; if you've invented a new safety feature for a horse saddle stirrup, well all I can say is at least US patents only go back to 1790.

In a couple more years Google Patents may revolutionize the patent search process.

There is hope that the entire U.S. patent database will soon be keyword searchable, by the way. Google has an ongoing project called "Google Patents" (http://www.google.com/patents) essentially comprising a search engine that builds its database by crawling through the pre-1976 patent images using optical character recognition. The tool is already fairly useful; but, the current state-of-the-art of optical character recognition being what it is, it misses things, and so to be thorough one must still do a US classification code search. But in a couple of years, who knows?

3.5 FINAL TIPS BEFORE YOU START YOUR SEARCH

While you peruse patents, take the opportunity to familiarize yourself with some of the field-specific jargon.

While you are doing your searches, there is one other bird you can kill with the same stone. This is a handy opportunity to familiarize yourself with the specific language of the field of your invention. There is often unique terminology related to a particular field, and these won't necessarily be ones you're familiar with, even if you are an expert in that field. In fact, some of the terminology really only comes up in patents. Don't be intimidated! There is no actual requirement that you use such terms, and no one really talks like this, including the author of the patent you're reading. Nevertheless, using the traditional language can be time saving, because often these terms convey a substantial amount of meaning when used in their appropriate context, which can save you half a paragraph of description.

While you conduct your searches, pick out three good examples to reference when you write your own.

While you are doing your search try to pick out about three patents that are particularly good and exemplary. Keeping these next to you as you write your patent will be time-saving and a sure cure for the occasional case of writer's block.

Do everything you can to minimize your emotional investment at least until after you've completed a thorough patent search.

As a final note, it is likely that you will experience some anxiety or a sinking feeling in your chest as you do your patent search. You'll be genuinely frightened with each click that your dream will suddenly be dashed to pieces. Honestly, this is because you've probably waited too long to do this search, and you're already too emotionally involved. You are, after all, only human. All I can do is to assure you that this is unfortunately quite normal, and that even if you find your invention already exists, EVERYTHING IS GOING TO BE OK. You'll have another in no time. Next time, the first thing you'll do is the evaluations discussed in the preceding chapters and a patent search before you become over-invested, right?

Not finding prior art does not guarantee it does not exist.

All you can do is be as thorough as possible.

No patent search can actually prove that no prior art exists for certain unless you actually look at every patent that has ever been issued (and there's still the open issue of prior art outside of the patent database). At an average rate of reviewing one patent a minute that would take well over thirteen years. You couldn't actually do such a search even if you wanted to, however, because most patents filed prior to Dec 15, 1836 were irreplaceably destroyed by fire. Obviously the destroyed patents were effectively cancelled, so you don't need to worry about them. Your present task is to convince yourself that something (prior art) doesn't exist. If you find prior art, then you can be 100% sure that your conclusion that prior art exists is correct, but the converse is not true, for if you don't find anything, all you know for certain is that you haven't found anything. Clearly, the more you search and don't find anything, the more probable it is that no prior art exists – so be thorough!

Good patent search documentation will be valuable down the line.

As you fill out your patent search log, try to make your notes clear such that they would be understandable to someone besides yourself (or yourself several months down the line). Ideal timing will have you pursuing license agreements during the patent pending period of your application. As such, prospective licensees will have no way to know for certain that your invention does not infringe prior art, but for you to be able to provide excellent documentation of a thorough and organized patent search will go a long way to assuage their fears. Later on, if your project does ultimately progress to the point where a licensee wishes to engage more involvement by a patent attorney than simple review of the patent application, your patent search documentation will be of great benefit to the attorney (and therefore your licensee's budget), and you'll stand to gain considerable respect.

If the first patent was issued in 1790, why is U.S. Patent No. 1 dated July 13, 1836?

Because Patent No. 1 isn't the first patent issued in the U.S. Originally, U.S. patents were not numbered, but issued simply by name and date. Numbers were introduced by the Patent Act of July 4, 1836. But, rather than retroactively assigning a number to the nearly 10,000 patents already existing at the time, it was decided to just start counting from that point forward, such that one Senator John Ruggles of Maine (who also happened to chair the "Committee on Patents and Patent Office" with undubitable conscienscious objectivity) received U.S. Patent No. 1 on July 13 of the same year. Ironically, the majority of U.S. patents were destroyed by fire just five months later. (Even more ironically, the same government that had never in forty-six years elected to make and store back-up copies at a separate location had recently begun construction of a fire-proof building for their safekeeping). Most of the lost documents were simply cancelled, but about a quarter were able to be restored from private records and assigned numbers beginning with "X"; thus, the first U.S. patent currently bears the designation X000001.

3.6 OK, LET'S GET STARTED.

Action Step 4

Perform a comprehensive search of the USPTO patent and published application databases for prior art relating to your invention. Record your searches and results in the worksheets provided in Section 1 of the Application Workbook. The steps you should take are summarized below:

Part I - Keyword Search:

1. Search from most specific to general. Record all searched expressions in the Patent Search Log in the Application Workbook (add lines as needed). You may find it quicker to first type search syntaxes in the log, and then cut-and-paste them into the search fields, instead of typing them twice. You can brainstorm a bunch and enter them on the list before you go the web. As you work, whenever something jogs your brain to think of another good keyword search, take moment to enter it in the next line before you forget it. Only fill in the "Date of Completion" box when you have actually finished a search, so you'll end up with a list that grows as you go; but you'll always know which searches you've done and which are still on your to-do list by the status of the completion date boxes.

2. Record any patents with similarities to your invention in the Search Results Table below the Search Log in the Application Workbook (again, just add lines as needed).

3. Write a short note regarding what is similar about each table entry.

4. Record the US classification code for each entry.

5. Remember to review each patent you enter into your table as you go to familiarize yourself with field-specific patent terminology.

6. Make sure you pick out three patents you want to use as table-side style references as you compose your own specification (on the worksheets highlight them or put an asterisk next to them).

Part II – US Classification Code Search

1 Review the classification codes for each item you entered in the Patent Search Results Table in the Application Workbook. Append each unique US classification Code as an entry in the Patent Search Log, starting beneath your last keyword entry. Don't search them yet – first go on to Step 2 below.

2 Search the US Classification index for relevant classification codes:

http://www.uspto.gov/go/classification/uspcindex/indextouspc.htm

Add each relevant code that your find not already appearing in your Patent Search Log at the bottom of your list. You now have your baseline set of search codes.

3 Search all patents (1790 to present) related to these classification codes.

4 Append any additional patents you find with similarities to your invention to the Search Results Table, with a short description of the similarities and US classification codes.

5 Add any new search codes you encounter to the bottom of your Patent Search Log, and search them as well.

3.7 WHEN YOU'RE FINISHED

If you approach the end of your patent search with the one surviving ambition that it just come to end, your patent search may, in fact, be thorough.

Finished? If you've applied yourself diligently to the method outlined in the steps, you have now completed a fairly rigorous patent search, and I expect your ready to do something, *anything*, else. There are no specific questions I can ask you to guarantee you've thought of everything, but, if you're pretty sick and tired of looking at patents hour after hour, night after night (if you work a day job), that's a good sign. Performing a good patent search is somewhat like beating your head against the wall. It's just about that entertaining, and it feels just as good when you finally get to stop. Hopefully your search, in fact, came up negative – that is, you didn't find anyone had already patented your idea, or at least you found that enough of your idea is original that you feel you can and want to proceed.

Your patent search was either:

Negative
(You didn't find disqualifying prior art)

Or

Positive
(You did find disqualifying prior art)

Congratulations! If you're idea is good, technically feasible, and original, not only am I starting to think that you might be genuinely smart, you're certainly lucky (people debate which is better, but, hey, when you have it all, who cares?). Very few people have even one original thought in a lifetime (there's just a lot of thinking going on, maybe not much on average per capita, but still a lot), but you may actually be onto something.

Still, let's be enthusiastic without getting overly optimistic – to make any money on your invention a number of things still need to fall into place, but you've made it this far, and you're ready to move on to Chapter 4

Look at the bright side – at least someone else thought your idea was pretty good, and believed in it enough to get a patent. This is the normal conclusion to a patent search, which is why I say do the search before you get too attached to the idea.

I said it before, and I'll say it again:

DO NOT BE DISCOURAGED.

Think of it this way – you've verified that you had a patentable idea. Whether or not some one else happened to think of it first is entirely out of your control, so you get an A in Inventorship 101. For you to have thought you might be the first inventor of your idea, and then found it in the prior art means it probably never made it big (or you'd expect to know about it if you're familiar with the field). Chances are, then, that the unlucky one is the guy who spent money on a patent that went nowhere (as previously discussed, that's most). In the grand scheme of things, that guy just saved you a lot of time and money, and you can move on to your next invention, that might really be a hit. Inventions really tend, very much, to be a numbers game. It's hard to completely predict what's going to be a market success (although sometimes it's pretty easy to spot a guaranteed flop), but every time you come up with one, you get another throw of the dice.

4 Making It Real

4.1 Prototyping

Prototyping and patenting go hand-in-hand. Without IP protection, you may get taken to the cleaners. Without a prototype, you may not get taken seriously.

Up to this point, it would have been very premature to do anything more than evaluate your idea, because prior to a thorough appraisal and patent search, you had no basis to justify investment of time and money into your idea. Presumably, if you're starting this chapter, you now do. At this point, you have justification to invest a little cash in your invention, but your decisions should still be highly value driven. I strongly recommend that you embark on the construction of a prototype, for the basic reason that you are going to find it exceedingly helpful in marketing your idea to prospective licensees or investors. I once had a guy approach me with an idea he had for an electric car. He had developed the idea as far as to give it a catchy name. Obviously, he held my attention as briefly as I could possibly manage without being downright rude. I additionally recommend that you at least commence prototyping before you write your patent because in the process of prototyping you will make discoveries that you will want to include in your patent claims.

Strictly speaking, prototyping is beside the primary objective of this guide, the specific intent of which is to help you create a patent and license your invention. Nevertheless, it is my belief that prototyping, writing a patent, and marketing go hand-in-hand. Generally your invention won't go far without both a patent and prototype, so it would be quite remiss of me to entirely focus your attention on the patent process and not even mention this equally key component to having a real chance to make your invention a success.

Start by defining a prototyping plan of feasible scope.

When we talk about making a prototype, there is no question that we're talking about a task encompassing a rather broad range of scope. Obviously inventions vary greatly in size, complexity, technical difficulty, and technical risk. I like to group inventions according to the following criteria:

Simple

Your invention may be one of those clever ideas that has been overlooked but anyone could easily make, which is indeed the best kind of invention of all. If your invention is simple to fabricate, well then, you have no excuse not to make a prototype, and thankfully it won't take very long.

Complex

Watch out if your invention comprises a large part count. Is it practical for you to build this? Consider whether or not you really need to build all of it to be convincing, especially if we're talking about something composed of repeated units, e.g. a bicycle chain. If I invented a new kind of bicycle chain, sure I'd want to make a complete chain, but I might find it more practical just to make several links for show.

High Technical Risk

How long do I think I'll need to fiddle with this before it works correctly? If your invention is a machine, let me assure you it'll never work perfectly the first time you try it. Are there elements to the invention that are unproven? Clearly, a machine is lower risk if it's comprised of components that all already exist and are used elsewhere, especially if they are commercially available off-the-shelf. Do not underestimate the time it will take to work out the bugs. You may have to make multiple prototypes, each correcting deficiencies of the last – this can take years. Conversely, it doesn't have to be absolutely perfect to move forward, but it is very difficult to market an invention (either to licensees or investors), if the remaining technical risk isn't considered pretty low. If your invention does require significant technical development, you, more than anyone else reading this guide, need to develop some form of in-house prototyping capability that will meet your needs. If you have to go out to professional machine shops for every little modification of every little part, costs will overwhelm you fast.

Just Plain Big

Most inventions are small enough to inexpensively prototype, but what if you've invented a new type of modular housing?...or a better tractor trailer?...or a more effective snow plow? Most inventors are not going to be in a position to shell out for something akin to these examples, even if we're talking just buying materials. Even if you could afford the materials, it's just not practical to build a submarine using just your evenings and weekends. This almost goes without saying, but, build a model. Clearly, it won't be as impressive as the real deal, but it'll be way better than just waving your arms.

Inherently Expensive	Even if an invention isn't enormous, it still may be very expensive to prototype. Again, here one has to be practical. When it comes to marketing your product, no one is going to expect you to have invested $100k working in your garage – occasionally you hear about someone who did that; but, if you show up in a board room and tell them you took out a second mortgage to build a demo unit, sure they'll know you're serious, but they'll also know you're a fool. Find a way to demo your concept without breaking the bank.
Non-Physical	Ok, if you've invented a method or a process, such as a new business practice that you intend to peddle as a consultant, you're not going to need a prototype. Nevertheless, you can't just show up empty handed. You'll need marketing materials, training manuals, etc. You still have a lot of work to do. Obviously the same goes for something relating to software – better at least program a demo.

Make your prototype yourself if at all possible.

Make your prototype yourself if at all possible. If you're creative enough to come up with an idea worth patenting, you should be able to find a means to get a demonstration prototype together without betting the farm. If you pay attention, you'll hear lots of stories about people who have spent tens of thousands having prototypes professionally fabricated and they show you a piece of plastic. As a minimum, always do the prototype design work yourself (we'll come to that next), even if you must engage the services of others for the actual manufacture.

Keep it simple.

Keep it simple – you're not actually creating the market product (unless it's so simple that's practical), just a representation of it. In particular, materials of construction do not have to be what you would actually see on the market – aluminum instead of steel, plastic instead of metal, etc. Clay, paper, styrofoam; it all looks the same under spray paint – use whatever will suit your immediate needs. Non-structural elements can be paper maché, if that'll do the trick. Does your invention need to be weather resistant? Even if it does, chances are your prototype doesn't. These days, there are dozens of inexpensive casting products available (but ordinary fuse beads flow fairly well at about 350 °F and don't begin to burn until about 400 °F, making them a reasonable source of plastic that can be melt-cast in air; hot glue is also a good option for moderately flexible parts), and, after all, you only need to make one prototype. Also, don't re-invent the wheel. This should go without saying, but I'll say it anyway for completeness – if you can buy something and convert it into your prototype, do it. Building a chainsaw with a special safety feature? – of course you shouldn't build a chainsaw from scratch. Modify one. Change only what you need.

The intent of your prototype is to convince, not complete.

If you're following my ramble, here's the concept that I'm trying to form in your head – when it comes to prototyping, anything goes. Generally, the purpose of your first prototype is to convince, not to complete. You're only objectives are that your prototype work, and look good enough to impress. How good? That depends. If you've invented a machine that turns lead into gold, well go ahead and just let form follow function. You'll get more attention than you want even if your device resembles three chickens strapped to a golf club. On the other hand, if you've accomplished the even more difficult task of inventing an ergonomically friendly office chair that doesn't feel like a straight jacket, I'd recommend making your prototype look pretty slick.

Never forget safety.

Cut whatever corners you think are in your best interests of saving time and money, but don't skimp on safety. It's just not worth it. Some bad things you can do are obviously stupidly dangerous to most people, many involving various cutting tools. What I find many people ignore are respiratory hazards, such as nasty solvents and fine particle generating materials. Take that stuff very seriously, and a particle/chemical vapor mask only costs about $30. Those disposable paper things are a joke that, frankly, shouldn't be on the shelves, by the way. If you're melting or cutting any type of plastic, or gluing practically anything other than paper or wood, keep the area well ventilated and wear an appropriate respirator. Don't forget eye protection.

4.2 Design – More Important Than You May Think

Design is both part of creating a patent application and the first step in reducing an invention to practice.

The beginning of prototyping is design, and you should not get this confused with invention. You invent the thing when you come up with the basic idea of how to solve a problem or meet a need. Design is the first step of reducing that idea to practice, and it applies both to the writing of a patent, where presentation of a workable embodiment of your invention (known as the "preferred embodiment") is a qualifying requirement, and creating a prototype. You may have a great invention, but if you cannot also come up with a decent design, your invention is probably sunk. Ultimately, a market product development process, whether by a licensee or your own enterprise, may supersede much of your original prototype and patent-stage design, but you need to do everything possible to make sure you get to that point.

The secret to good design is refine, refine, refine.

Do not stop with the first design that you come up with. Once you have a concept, plan to spend many hours refining it and thinking of ways to better it. You cannot expect to preempt all future possible design improvements, but you should certainly strive to pick all low-hanging fruit. If a third party can spot an improvement after looking at your design for ten minutes, you haven't done your homework – and it weakens your patent. Technically, it is the "CLAIMS" in your patent that define the scope of your intellectual property, not the example embodiment(s). Nevertheless, you can't claim what you haven't figured out yet, and all of those improvements you've missed are potentially patentable by someone else. If your claims are well drawn they won't be able to use the improved design without obtaining a license from you, but your path to a practical embodiment may be equally blocked by them.

> A patent with a weak preferred embodiment is a weak patent indeed.

A bad initial design can be very hard to overcome.

In fact, I would go as far as to say that if you first lay out a conceptual design and then fail to find significant improvements that make your original design seem fairly sad, there's a 99% chance you've missed something important. I've seen many patent embodiments where it was very clear after just perusing the drawings for a few minutes that little refinement had gone into the design. I recall one case where the patent embodiment was blatantly just plain bad, had at least 30% more parts than really were needed to accomplish the function, and was pursued all the way to market with very little improvement. The initial product was awful, and was only sustained on the market because the inventor/entrepreneur could afford to throw gobs of money at it until ultimately the product evolved into something successful (bearing little resemblance to the original market entry) through good design work by third parties (who ultimately purchased the original patent rights). If throwing copious sums of cash at a half-baked idea sounds good to you, may I recommend running for congress where it seems you might fit right in (and it won't even be your own money you're throwing around!). In the case of the design of a practical embodiment for your invention, you'll find that a stitch in time really does save nine, not to mention your sanity when it comes to fabricating your prototype.

4.3 Getting Down to It

To make a prototype, you'll need time, tools, materials, and skills.

Ok, enough said. You will need four things to create a prototype: time, tools, materials, and skills. Preferably, your invention pertains to a field where you already have personal knowledge and experience, that is, you are what is referred to in patent lingo as one who is "skilled in the art". Now, having come this far, I'm inclined to think you're probably pretty smart and have some drive, and so none of these prerequisites is beyond your reach if you want them. You can manage your time, you can obtain the tooling you need (having worked your prototype design to the point where it can be accomplished with affordable tools and materials), and you can learn what you don't know (you're doing that right now; and, these days, it's amazing what you can learn on the internet).

The amount of time you have will usually be the largest factor determining how long it takes you to complete a prototype.

Be realistic about your timelines. Limitations on time, tooling, and existing skills will do one thing for sure, and that is protract the prototyping process. If that's not a problem, and you prefer to work alone, go for it. I myself have never made a prototype without enlisting the help of a partner. This is for two reasons. Firstly, I'm a busy guy, and I don't typically want to invest the required amount of time into a single project (I have two other projects in the works at the same time as I'm writing this guide, for instance, in addition to my full-time gig as a propulsion engineer, husband, and dad). Secondly, I have an excellent friend with whom I like to work.

Taking on a partner with a complementary skill set to your own can greatly facilitate progress.

Do not underestimate the value of taking on a partner. Selection of the right partner can greatly enhance your probability of success, both by multiplying the rate at which your project can progress, and augmenting any deficiencies you might have in regard to the requirements of your particular product. Not good with your hands and need to fabricate something? Not a proficient computer programmer but need to demo a new concept for an operating system user interface? (Software industry patents seldom have anything to do with specific code, as algorithms, like mathematical equations, aren't patentable, by the way.) Just plain don't know a thing about robotics, but need a body for that positronic brain you've just invented? Get someone who is, does, whatever.

You may find more perseverance in a partnership than solitude.

There is a secondary advantage to taking on a partner and that is companionship. You're going to spend considerable time on your invention, and honestly, for most people that's going to get pretty lonely before they've seen it through. Not everyone, but most people. (There might be an exception for computer programmers, whom I've always regarded among all engineers as probably the most comfortable in the presence of the dead, or at least the undead, and, interestingly, they do keep similar hours.) Most people simply will find it easier to persevere with a partner than without. The Wright brothers. Proctor and Gamble. Bill Gates and Paul Allen. Bonnie and Clyde...even criminals usually find someone like-minded to work with.

Taking on a bad partner would likely spell doom for your project.

If you do take on a partner, choose carefully.

A good partner is worth his or her weight in gold. Unfortunately, a bad one is worth the same amount chained to your ankle on a sinking ship. Be exceedingly careful in selection of a partner if you elect to do so. Look specifically for someone who has a balance of time, skills, and tooling that complements your own. The closer they are to you, the better (provided they have the above). A spouse, a teenage offspring (if you have one who still listens to you), a sibling, or that college friend with whom you are inseparable make good choices if your relationship is such that partnering is possible. Above all, get someone you know well and can trust implicitly. What you need to know most about them is that they'll do their share and won't quit on you. Don't have anyone like that? Well, I guess you're on your own (but you really should be looking to fix that). At least you won't have to split the pot, should you ever make anything. By the way, patents are inheritable property, so doing away with your partner won't generally do you any good.

Do not wait until after you have completed your prototype to start writing your patent application.

You do not have to fully complete your prototype before moving on to the next chapter, which is where you are actually going to start writing your patent. Writing a patent application is not the same as filing for a patent. Your can write it, and then later make critical decisions regarding if and when to file it, but you certainly can't file it before writing it. Composing the patent document itself can be a big job, so don't procrastinate, and you may find it more pleasant to you spread the task out a bit, working in parallel with your prototype development. Remember, as I discussed in Section 1.7, when it is best to file your patent in terms of balancing risk that someone else will file something similar against maximizing your patent's life and delaying expenses is a judgment call that is very field specific, and one you will have to make for yourself.

So, without further ado:

> **Action Step 5**
>
> Carefully plan and then execute the fabrication of a prototype. Remember, extra hours in planning and design stage can save you weeks later.
>
> Meanwhile, continue on to the next chapter.

The Inventor Who Became President

Did you know the only U.S. President to ever receive a patent was none other than Abraham Lincoln? Lincoln displayed a lifelong fascination with mechanical things, often pausing to study tools and machines he chanced to encounter. Homeward bound between sessions of Congress in 1848, Lincoln observed the use of barrels and other makeshift means of buoyancy to dislodge a flatboat upon which he was traveling from a sandbar. The following year he received Patent No. 6469 for a "new and improved manner of combining adjustable buoyant air chambers with a steam boat or other vessel for the purpose of enabling their draught of water to be readily lessened to enable them to pass over bars, or through shallow water, without discharging their cargoes..." Lincoln's idea wasn't terribly practical, and as somewhat of a relief to friend, law partner, and skeptic William H. Herndon with whom Lincoln often shared his plans for a revolution in river travel "the invention was never applied to any vessel, so far as I ever learned, and the threatened revolution in steamboat architecture and navigation never came to pass."

A scale model made by Lincoln himself is on display at the Smithsonian Institution in Washington, D.C. If you're curious you can find a photo of it at Abraham Lincoln Online.org:

http://showcase.netins.net/web/creative/lincoln/education/patent.htm

Starting the Application

5.1 WHAT YOU WILL NEED

How's that prototype coming? OK, I invited you to read ahead, so the first time you read this you probably won't haven't started it yet, but do get that going. So it's time to dig in – with this short chapter we will kick off work on your patent application itself. Let's start by going over what you will need.

A Computer

If you've completed Chapters 2 and 3, I'm thinking you must have access to a computer. The truth is that, these days, a brand new computer will be quite a bit less than the patent application fees. It's not that the fees are all that bad, but computers have become very cheap. You won't need anything fancy. There's really nothing involved that that couldn't be done pretty much just as well with the technology available twenty years ago.

A Printer

Again, you won't need anything high tech (but if we're talking dot matrix, it's time to upgrade). The USPTO lists "permanent black ink", but I believe that's in a context where permanent means non-erasable, not necessarily waterproof (e.g. a laser printer, which are pretty cheap nowadays, by the way). Technically, you won't need to print until you're all done and ready to drop your application in the mail, but I find it's useful to be able to make hardcopies of the drawings to look at while writing.

Word Processing Software

I've tailored this document to Microsoft Word because it's by far the most standard, and so that I can illustrate some special techniques by examples, but you can pretty much use anything you want. Any decent word processor will support those techniques, although obviously software-specific details of implementation will vary. The Patent Application Workbook is in MS Word format, but most everything these days should be able to open it. Also, let me assure you that MS Office hasn't appreciably improved since 1997, (actually most people seem to think it's become steadily worse) and that embarrassingly lame picture placement routine is as bad as ever, so don't worry about upgrading if you haven't done so in a while. You're not missing out on anything.

In fact, I strongly recommend that you not upgrade. (My apologies Microsoft, but you know exactly what I'm talking about.) I'm certain that most at this point are already painfully aware that Microsoft made significant architectural changes to MS Office in 2007, which left the world somewhat perplexed. Menus and toolbars were replaced with a single multi-tab "ribbon", which seems to eliminate the advantages of both. The grapevine has it that the development team excluded from their focus groups anyone who had previously used MS Office, which would explain a lot. At any rate, as a long-time power user of all of the MS Office applications who is proficient with both the old and new user interfaces, I can confidently report that the new version is quite a step backwards, and moreover is absolutely rancid with some pretty obstructive bugs and feature omissions. If you need to procure a copy of MS Word for your project, I really do recommend trying to obtain a pre-2007 version, and ideally pre-XP (the going rate is about $10 on Ebay).

If you are working with MS Word 2007/2010, don't fret – we can make it work and all the procedural instructions in this edition address both the old and new versions. In spite of all, I still do recommend working with MS Word, not because it's particularly good software (which it's not), but simply for compatibility with others (such as attorneys) who will likely eventually become involved in your project. But wouldn't it be nice if the MS Office developers could get it together?

A Drawing Program

Unless your invention falls into the non-physical category, you'll need to make drawings. You could do your drawings by hand, but I really don't recommend this. You will find that hand drawings come with the significant disadvantage that they can't really be edited after they're complete – so if you have to make a change (and you will), you'll end up needing to do the whole thing over. Moreover, often you will have a number of drawings that represent minor variations on a theme. Electronic drawings make this easy, since you can just copy/paste/edit; but if you choose to work with hand drawings, you're on the hook to recreate the entire drawing for every variation. Even if you've never tinkered with a drawing program before, you're going to find them a big time-saver.

Depending on the complexity of your invention, your software needs will vary, but you definitely don't need anything super fancy. In many cases, the MS Office drawing tools will be sufficient, but they do have two significant shortcomings in that they won't do Booleans (i.e. allow you to make complex objects by welding shapes together or cutting one out of another), and they can't really do cross-hatching (the texture-based cross-hatches are actually fairly low resolution bit-maps, which look pretty lame and you only have a few angles to choose from). In fact, you will find that how you choose to do cross-hatching will be a key discriminator with regard to how you proceed to create drawings. You can use a drawing program that has a vector-based (not bitmap) hatching tool. The trouble you'll run into is that no drawing tool that I am aware of correctly exports cross-hatching to a vector-based image format compatible with MS Word (e.g. WMF, EMF), which means that if you do hatching in your drawing program, you'll have to convert your images to some type of pixel-based format (PNG is generally the most compact) to get them into MS Word. For good image quality on a graphic made of up thin lines (such as a typical patent drawing), you'll want at least 300 dpi. That will work fine, and, since the images are black and white, won't really be too large.

But it's frustrating – there's something genuinely silly about converting a perfect vector-based graphic into any kind of bitmap. Because vector-based graphics store the data as objects – lines, shapes, characters, etc., the file size is tiny, but even still you get maximum resolution from the printer, since the driver effectively tells it exactly what you want. Conversely, the pixel-based graphic (BMP, TIF, PNG, JPG, and the like) unnecessarily takes up much more file space and only approximates the resolution recorded in the original line drawing. As such, I have created a library of cross-hatch WMF's which allows me to do the cross-hatching in MS Word after I have imported drawings created in another program. (I first import my drawings as WMF's into MS Word, convert them to MS drawing objects, and then embed the cross-hatch as fills). While this involved some extra work up front, for patents with many images this has really kept the file size down, and since embedded images can be copied from one object to the next with the format painter, it's really not that bad at all. Be warned, however, that bugs in the 2007 and later versions of MS Word effectively destroy this option completely – if you're using one of those versions, you'll be stuck converting to bitmaps.

There are a number of freeware drawing programs out there that should do the trick if you don't already own one you like. Before you put any time into a drawing, make sure the program has a decent cross-hatching feature if you intend to use it for such. So what do I think is good? Try Inkscape.

Time

How long it takes to write draft patent specification and drawings will vary substantially based on the scope of the invention, but do not underestimate the level of required effort. I continue to emphasize this because there is only one thing I can think of that is worse than doing nothing with a really good idea, and that's getting halfway there and quitting. People start big undertakings out of interest and excitement, but they finish them only for the simple reason that they are in the habit of finishing what they start. The glamour will be long gone before you are done with your application, leaving you with only a choice between apathy or determination. Count the cost at the beginning, but once you proceed, persevere.

Writing Skills

Technically you don't have to be a very good writer to create a patent application (but it sure makes it easier), and this guide is going to help focus your efforts on what is important. You do need to have the ability to put your thoughts down on paper in an understandable fashion. Nice prose won't do anything to strengthen your patent, but it is important to be clear. If you are in the unfortunate position of having fairly poor writing skills, this would be another good reason to seek a partner whose strengths compliment your weaknesses.

5.2 GENERAL REQUIREMENTS TO APPLY FOR A PATENT

Your patent application must (and will upon completion of this workbook) include the following:

The elements of a patent application

- Utility Patent Application Transmittal Form (or Transmittal Letter, but we'll just use the form)
- Fee Transmittal Form and Fees
- Application Data Sheet (see 37 CFR § 1.76)
- Patent Specification and CLAIMS
- Drawings (Unless not applicable)
- An Abstract
- Oath or Declaration
- Nucleotide and/or Amino Acid Sequence Listing (If applicable – can't say that I'm an expert with the specifics of this one.)

That may seem like a lot of stuff, but the fact is that they're all very short and simple with the exception of the patent specification and drawings, which are where, by far, the bulk of your effort will lie, and where we will start.

Let's start with a few basic ground rules:

1 Your application must either be in English or be accompanied by a translation into English to be accompanied by a statement of accuracy per 37 CFR §1.17(i). Australian will do in a pinch (wink). Obviously, since you're reading this guide in English, I'm assuming you also have the capability to complete the worksheets in English.

2 The page size must be either:
- 8½ × 11 inches (21.6 × 27.9 cm)

  or
- 21.0 × 29.7 cm (DIN size A4).

3 The left margin must be at least 1 in (2.5 cm). The top, bottom, and right margins must be at least ¾ inch (2.0 cm). Drawings have other rules which we'll discuss when we come to Chapter 6.

4 Lines must be either 1.5 or double-spaced. The Application Workbook is pre-formatted to 1.5 line spacing.

The Application Workbook is preformatted to make life easy.

Generally, you won't have to worry too much about formatting, because naturally I have pre-formatted the Application Workbook sections for you. For completeness, I'll mention that the USPTO requires your specification to be printed in black indelible ink (no hand-writing, but you could use a typewriter if you're really retro), single sided, portrait orientation, on white paper that is "flexible, strong, smooth, nonshiny, durable, and without holes." All pages must be of the same size.

5.3 The Parts of a Patent

Your application will comprise an abstract, specification, and drawings.

The parts of the patent application that will become your patent are the abstract, specification (including and especially the claims), and drawings. As you may expect, the abstract is a quick blurb that goes on the front to aid people in quickly assessing the patent's contents. The specification presents the who/what/where of your invention, basically teaching the invention, stating your claim(s), and providing a basis of proof of patentability; that is, that your invention is novel, useful, and non-obvious. As such, the specification is comprised of:

The elements of a patent specification
- The Title
- A list of inventors
- A section of text describing the general background of the invention
- A brief summary of the invention
- A detailed description of the invention including at least one example embodiment
- A list of claims (the most important part)

Expect to jump back-and-forth between sections. Writing a patent specification is not linear.

We will work these sections one at a time in the Patent Application Workbook. Once finished, the first draft of your patent specification will be complete. Instructions will be provided as we go, but, once again, you may find it useful to read through the worksheets in their entirety to first get a global view of where we are headed. This is the order these sections should appear in your application, and so this is the way you'll find the worksheets arranged so that they'll be in the right order when you're finished, but this is not the order you will be working them. I'll guide you to each next section as we go, but you should also expect to jump back and forth between sections often, because, as you work, ideas about what you would like to put into other sections will be sparked by your progress in different areas (they all do, after all, relate to the same invention). When something occurs to you that will be useful in another section, don't expect to remember it! Jump over to that section and type at least a note to remind yourself.

It's much easier to do drawings first.

We will start, however, with the drawings, or, as it is said in patentese, the "several views of the drawing". You will find your writing efforts to be much more focused and efficient if you first complete at least draft versions of your drawings; which brings us to the next chapter.

6

Drawings

6.1 Drawing Rule Enforcement

If anything that forms a novel part of your invention can be drawn, it must be.

Since they say a picture is worth a thousand words, almost all patents have drawings by requirement. Wherever applicable, anything you wish to claim and any feature mentioned in your claims must appear in your drawings. The few exceptions, according to the USPTO, are "compositions of matter or processes", and even still, you are encouraged to think hard about whether or not drawings may facilitate understanding in these cases.

The USPTO is no longer a stickler about the traditional drawing format rules, but if you choose to ignore them, you do so at your own risk.

The USPTO provides very specific guidelines for formatting drawings; among these are some outright rules, along with some that function more as recommendations in that not following them will usually not raise objections by your patent examiner. Not all that long ago, all of these rules were strictly enforced, and there existed an entire industry of drafters who specialized in patent drawings. Eventually the USPTO made a conscious decision to relax enforcement of the requirements, which effectively both cut their processing costs, and expenses of the would-be patenter. Naturally this is particularly beneficial to the small independent inventor, where the fees of specialized professionals represents quite an obstacle. What remains important is that the drawings be clear and understandable. Always keep in mind that the USPTO's drawing guidelines were created for exactly this purpose, and the effectiveness of your patent depends upon it. The more recent drawing rule enforcement policies are not unlike the USPTO's modern attitude toward the non-obvious. For practical reasons, the USPTO will equally allow you to patent a bad invention or a patent with deficient drawings, but, should you end up in court, you'll ultimately pay the price for those shortcomings. So, while you will generally be given grace to ignore the drawing rules, it's a good idea to follow them as closely as possible.

It's still a good idea to follow the traditional drawing rules as best you can.

6.2 Learning by Example

We're going to review the USPTO's stated (on their website) drawing guidelines in detail, but initially I will simply provide and describe for you a simple example set of drawings, with the aim to first instill you with an intuitive understanding of what you're shooting for. While my intent is to keep it simple, most users of this guide will actually find every drawing feature they need demonstrated.

Study the example drawing set on the next page (Fig. 6-1). Illustrated is a notional invention for a "PASTRY THAT CAN BE EATEN WITHOUT GETTING STICKY FINGERS", A.K.A., the doughnut-on-a-stick. The example may seem silly, and really pushes the limits of non-obviousness, but technically one could probably actually successfully file for this (assuming someone hasn't already) – the real reason not to is simply that it's worthless. Don't get bogged down with what is illustrated, pay attention to how it is illustrated – if you make your drawings look just like this (minus the notes in italics), you'll have no problems.

Don't get caught up looking at the example invention. Look at the drafting.

Avoid Photos

The drawings are, in fact, drawings and not photos. Generally, photos are only allowed in cases where it is impossible to convey understanding of the invention through drawings. In my mind that means almost never, because I can think of little you can photograph that can't be drawn. Scanning the patent literature, you certainly won't find many photos, except when it comes to plants (recall that human-developed plant hybrids are patentable), where the subtle differences between one plant and its relatives truly would be very difficult to capture in a black-and-white drawing. If your invention is a plant or otherwise qualifies for this rather rare exemption, you will have found evidence of this in similar patents during your patent search. If you haven't (and I expect you haven't), well, there you go.

Photos are allowed, but you should avoid them.

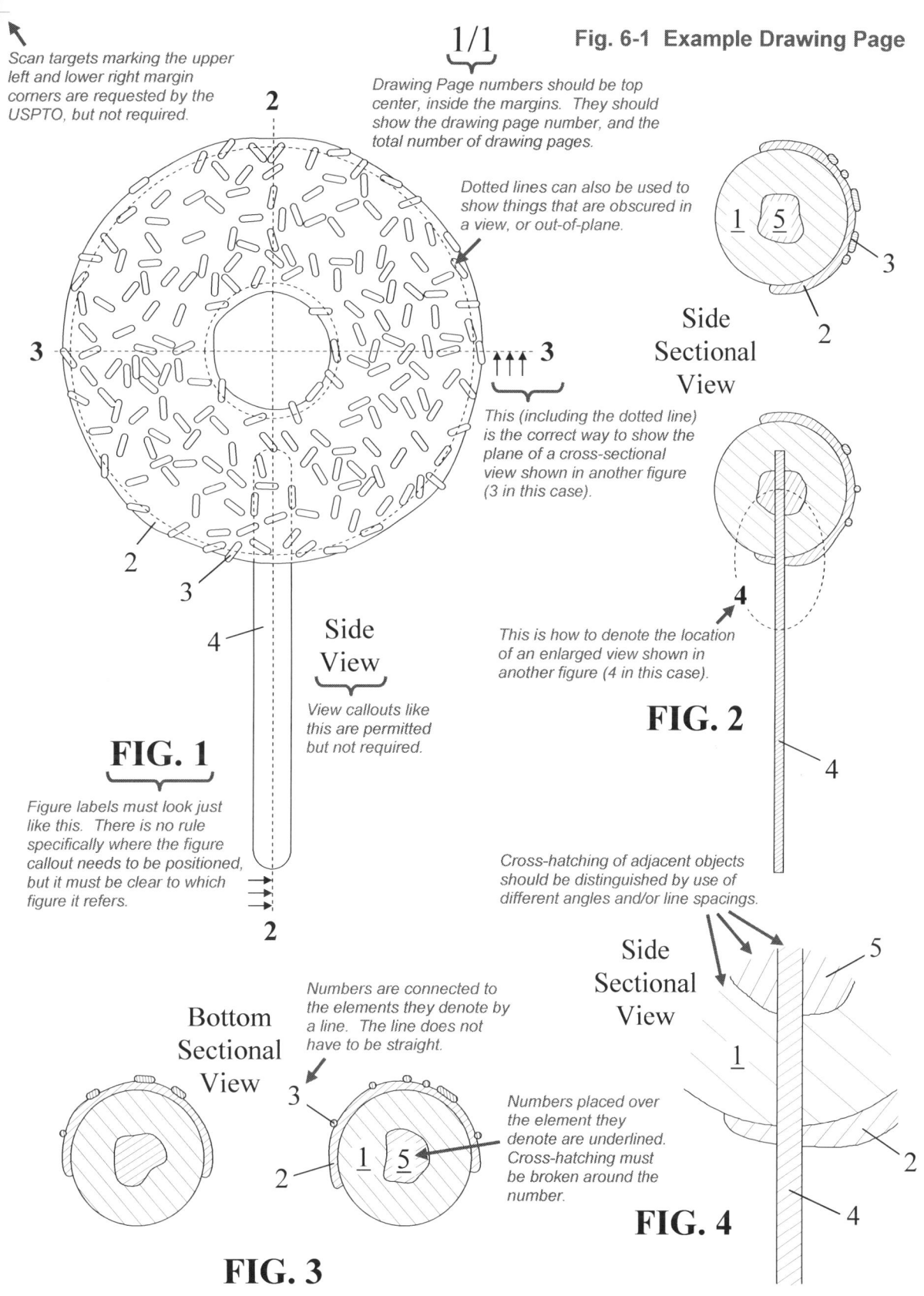

Fig. 6-1 Example Drawing Page

Black and White Line Drawings Are Preferred

Gray-scale drawings are less reproducible than straight black-and-white.

The drawings are black and white...and I mean black and white – no gray. Technically grayscale is allowed provided the USPTO can reproduce the images clearly (but you have no good way to verify that they do in advance), but not preferred – let me recommend not going there. Also, color is allowed only in cases where it is absolutely necessary. (Can you guess? Yup,...plants again.) As was the case for photos, I don't think one can come up with an example where one absolutely could not make a drawing clear using black and white, and special filing provisions apply if you do insist on color; so again, just don't do it. Black and white.

Each FIG on a Page Must Be Completely Separate from Any Others

Each figure can show only one view. No lines can go between figures.

The various FIGs share no lines, and there are no lines interconnecting them to show how the views relate. Each FIG is stand-alone and shows only a single view of the embodiment. You should place as many FIGs as will fit on one sheet of paper (keeping them in the correct order), but

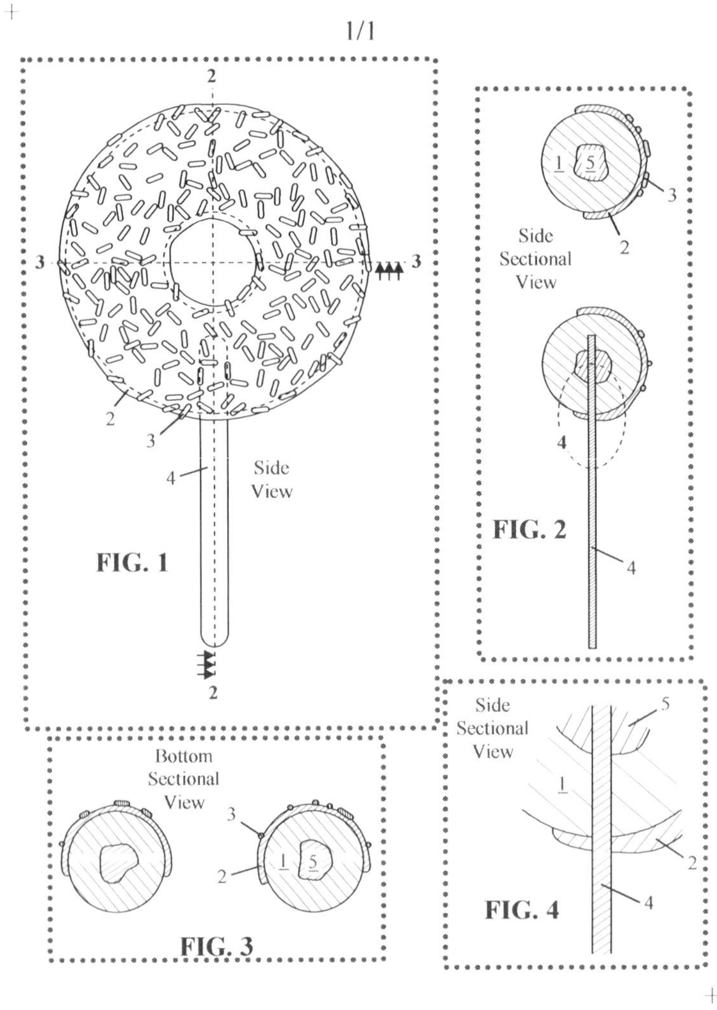

Fig. 6-2 You must be able to draw boxes around every figure on a page without any intersecting.

they must not intrude upon one-another. By this I mean that you must be able to draw a box around each of your FIGS, text and all, such that no two boxes intersect, as illustrated in Fig. 6-2.

Sectional Views

The location of sectional views should be denoted on unsectioned views.

All sectional views are indicated by a dotted line over another figure showing where the illustrated object is to be cut to produce the sectional view. At either end of the dotted line should be included the FIG number of the section view, but only the actual number. Use bold type to avoid confusion of this number with part labels. The dotted line does not have to be straight (although this is preferred), but may follow some natural path if it truly makes the sectional view more illustrative. Arrows must point at the side of the cut-plane that is depicted in the sectional view.

Sometimes you can get away without showing the plane of a sectional view on another FIG, but this is not recommended.

You are not required to have the location of all sectional views shown in your drawings, as often it is obvious. You should always show it if there is a figure where you can, but, as you browse the art, you will notice that very commonly no external view is provided, i.e. only a sectional view is shown since, in many cases, the external view is unimportant and interpretation of the sectional view is readily apparent. I recommend you steer a wide birth by always providing an external view that provides a frame of reference for any sectional view, but it really is up to you. Having the patent examiner come back and ask you to add it is a headache you don't need, though. There are cases where most patent examiners will insist upon having sectional views' planes clearly defined in another view, such as:

- Where you already have need of a view of the exterior that is intersected by the cut plane of a sectional view
- Where it really isn't obvious how the interior relates to an exterior
- Where objects in the sectional view interact with external components
- Where sectional views at more than one location are presented

Enlarged Views

Enlarged views should always be denoted on a larger view.

Close-up or magnified views of subsections of the invention should be shown outlined with a dotted line on a more complete view with the line intersecting the FIG number of the close-up view. Again, use bold type to make it easy to differentiate this number from part labels.

Drawing Text

Use of text in patent drawings is very limited.

No text should be present other than what is shown in non-italics in Fig. 6-1. You must include number labels for all parts, and the FIG numbers for any defined close-ups or sectional views. You must include the FIG label, exactly as shown, with no verbal title or description. The USPTO requires all characters to be ≥0.32 cm (1/8 inch). You might be able to get away with a 12 pt font, which is very marginal, but I have always used a minimum of 14 pt with good success, and recommend you do the same. Use 14 pt for everything except the FIG labels, where I recommend 18 pt bold. The view description labels, e.g. "Side Sectional View" are allowed but entirely optional – most authors don't bother.

Some exceptions to the standard drawing text rules are allowed, but you should really avoid them.

OK, there are some exceptions. Per § 1.84 (d) (listed in the next section), mathematical formulae, graphical chemical formulae, tables, and waveforms (data or mathematical function plots) are allowed, but honestly I've so far never actually seen one in practice where it was really necessary, with the exception of chemical formulae. The reason is that mathematical formulae, tables, and waveforms are typically only useful to provide credibility (which goes toward establishing usefulness) – they don't specifically address any other patent requirement, so they usually are needed only if the invention is outright unbelievable. Even if your invention's credibility is challenged, generally simply submitting a declaration that you have reduced it to practice will suffice. So, unless you're claiming the invention of a matter transmutation device or something else so far out on the bleeding edge of technological capability that you won't easily be believed, this type of corroborating material isn't necessary.

Legends are also one exception (§ 1.84 (o)), but words are to be kept to a minimum. Typically there's no need for a legend, since every element of something like a schematic is numbered and described in the specification.

By the way, you will find many examples of issued patents where authors have disregarded the text rules, but do not take that to mean they are correct to do so. As I alluded at the beginning of this section, patent examiners no longer enforce all of the drawing rules, but that does not mean it's a good idea to disregard them.

Cross-Hatching

Other than those used to depict the location of sectional and magnified views, dotted lines are not allowed except to:

Use of dotted lines is limited and must follow specific conventions.

- Show the location of an object that is not technically visible in the view shown (something that is under something else or above a sectional view plane).

- Depict an alternate location of a moving part. If this is not very simple and visually easy to interpret, you should instead opt to just show a second FIG that is identical to the first, except with the moving parts in their alternate locations.

All of the dotted line types designated by USPTO conventions are available in the MS Office drawing tools.

- Connect between parts to show how they are to be assembled in an exploded view.

No centerlines or anything like that. These days, much of the time, the patent examiner will not make a big stink about specific types of dotted lines for these purposes, but there is a convention, and you should follow it to be on the safe side:

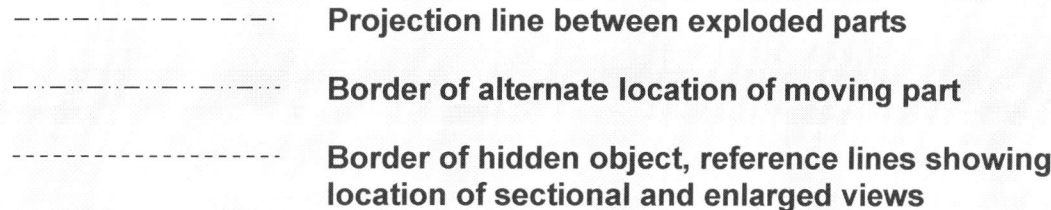

Projection line between exploded parts

Border of alternate location of moving part

Border of hidden object, reference lines showing location of sectional and enlarged views

All sectioned parts must be cross-hatched where they intersect the cut-plane.

All sectioned parts are cross-hatched. Often in a cross sectional view, a single part will not appear as a continuous manifold, because that part does not uninterruptedly intersect the plane – the part will look like a bunch of islands. All of these islands must be cross-hatched identically, or they will appear to be separate parts. The USPTO guidelines state that cross-hatching of different adjacent parts should be at substantially different angles to clearly distinguish them. Because pieces of a single part may appear totally disconnected in a sectional view, I recommend that you try to make sure that the cross-hatching of no two parts look the same. You don't have only the angle of the lines to work with, but also the spacing. You will find that it works best if bigger parts have larger spacing between hatches than smaller parts. No crisscross or double-line hatching is allowed.

Generally, you should try to make sure that no cross-hatched object or part of an object has fewer than three cross-hatches, or else the cross-hatches tend to be confused with drawing feature lines. For the same reason, cross-hatch lines should never be much further apart than ~1/8 inch. Also, even if a part is large, if it has small features, such as webs, this should lead you to use a finer cross-hatch.

Cross-hatching must be blanked out around text.

If you place a number label over a cross-hatched part (instead of connecting the number to it with a line), the cross hatches must stop around the label so that it is readable. Don't worry, this is easy since we will include all of the numbers as paste-overs in the drawing. When you create your cross-hatching, you won't have to worry about leaving spaces where you think numbers will go.

6.3 DETAILED LISTING OF DRAWING RULES

OK. If you just make your drawings the same as the example, you'll run into no problems, and that should cover 95% of the users of this guide. However, for those who have more esoteric needs, the following table includes the complete set of drawing guidelines provided by the USPTO on their website, with some useful commentary. In particular, I'll try to point out which of these guidelines are typically treated as rules by patent examiners, and which serve more as recommended practices (but don't get me wrong – I generally advise you to adhere to them all).

§ 1.84 Standards for Drawings

(a) **Drawings.** There are two acceptable categories for presenting drawings in utility and design patent applications:

(1) **Black ink.** Black and white drawings are normally required. India ink, or its equivalent that secures solid black lines, must be used for drawings, or

> Stick with this, no grayscale.

(2) **Color.** On rare occasions, color drawings may be necessary as the only practical medium by which to disclose the subject matter sought to be patented in a utility or design patent application or the subject matter of a statutory invention registration. The color drawings must be of sufficient quality such that all details in the drawings are reproducible in black and white in the printed patent. Color drawings are not permitted in international applications (see PCT Rule 11.13), or in an application, or copy thereof, submitted under the Office electronic filing system. The Office will accept color drawings in utility or design patent applications and statutory invention registrations only after granting a petition filed under this paragraph explaining why the color drawings are necessary. Any such petition must include the following:

> I've included this for thoroughness, but note the "On rare occasions". As I said above, color drawings are generally more trouble than they're worth.

(i) The fee set forth in § 1.17(h);

(ii) Three sets of color drawings; and

(iii) An amendment to the specification to insert (unless the specification contains or has been previously amended to contain) the following language as the first paragraph of the brief description of the drawings:

> The patent or application file contains at least one drawing executed in color. Copies of this patent or patent application publication with color drawing(s) will be provided by the Office upon request and payment of the necessary fee.

(b) **Photographs.**

   (1) **Black and white.** Photographs, including photocopies of photographs, are not ordinarily permitted in utility and design patent applications. The Office will accept photographs in utility and design patent applications, however, if photographs are the only practicable medium for illustrating the claimed invention. For example, photographs or photomicrographs of: electrophoresis gels, blots (e.g., immuno-logical, western, southern, and northern), autoradiographs, cell cultures (stained and unstained), histological tissue cross sections (stained and unstained), animals, plants, in vivo imaging, thin layer chromatography plates, crystalline structures, and, in a design patent application, ornamental effects, are acceptable. If the subject matter of the application admits of illustration by a drawing, the examiner may require a drawing in place of the photograph. The photographs must be of sufficient quality so that all details in the photographs are reproducible in the printed patent.

   "Not ordinarily permitted" should be warning enough. Unless you regard your circumstances as extraordinary, stick with drawings

   (2) **Color photographs.** Color photographs will be accepted in utility and design patent applications if the conditions for accepting color drawings and black and white photographs have been satisfied. See paragraphs (a)(2) and (b)(1) of this section.

   If you both regard your situation as a "rare occasion" and want to take on "not ordinarily permitted", have at it. Otherwise, think about how you can use black and white drawings.

(c) **Identification of drawings.** Identifying indicia, if provided, should include the title of the invention, inventor's name, and application number, or docket number (if any) if an application number has not been assigned to the application. If this information is provided, it must be placed on the front of each sheet and centered within the top margin.

   This will be taken care of by including your drawings with the rest of your filing, which has a section for this (see Chapter 10).

(d) **Graphic forms in drawings.** Chemical or mathematical formulae, tables, and waveforms may be submitted as drawings and are subject to the same requirements as drawings. Each chemical or mathematical formula must be

   As discussed above, unless your invention is so incredible that no one will believe it works without your providing test data or mathematical derivations proving it, you generally won't need to deal with this.

labeled as a separate figure, using brackets when necessary, to show that information is properly integrated. Each group of waveforms must be presented as a single figure, using a common vertical axis with time extending along the horizontal axis. Each individual waveform discussed in the specification must be identified with a separate letter designation adjacent to the vertical axis.

(e) **Type of paper.** Drawings submitted to the Office must be made on paper which is flexible, strong, white, smooth, non-shiny, and durable. All sheets must be reasonably free from cracks, creases, and folds. Only one side of the sheet may be used for the drawing. Each sheet must be reasonably free from erasures and must be free from alterations, overwritings, and interlineations. Photographs must be developed on paper meeting the sheet-size requirements of paragraph (f) of this section and the margin requirements of paragraph (g) of this section. See paragraph (b) of this section for other requirements for photographs.

> These days, this basically just means regular printer paper, either U.S. or European standard (wasn't the U.S. supposed to have converted to metric in the 60's?).

(f) **Size of paper.** All drawing sheets in an application must be the same size. One of the shorter sides of the sheet is regarded as its top. The size of the sheets on which drawings are made must be:

(1) 21.0 cm. by 29.7 cm. (DIN size A4), or

(2) 21.6 cm. by 27.9 cm. (8 1/2 by 11 inches).

(g) **Margins.** The sheets must not contain frames around the sight (i.e., the usable surface), but should have scan target points (i.e., cross-hairs) printed on two cattycorner margin corners. Each sheet must include a top margin of at least 2.5 cm. (1 inch), a left side margin of at least 2.5 cm. (1 inch), a right side margin of at least 1.5 cm. (5/8 inch), and a bottom margin of at least 1.0 cm. (3/8 inch), thereby leaving a sight no greater than 17.0 cm. by 26.2 cm. on 21.0 cm. by 29.7 cm. (DIN size A4) drawing sheets, and a sight no greater than 17.6 cm. by 24.4 cm. (6 15/16 by 9 5/8 inches) on 21.6 cm. by 27.9 cm. (8 1/2 by 11 inch) drawing sheets.

> Stay within the allowed margins – you are not likely to be forgiven this transgression by your patent examiner. I prefer one-inch margins all the way around to keep it simple, and the Application Workbook is already formatted for this.
>
> Certainly you won't find many patents that actually have scan targets on the drawing pages, and, frankly, I have to admit I've never included them. (Either I never noticed this before, or simply forgot about it.) Anyway, if the USPTO finds these helpful, why not oblige? You'll find them in the example and pre-formatted in the Application Workbook.

Detailed Listing of Drawing Rules 57

(h) **Views.** The drawing must contain as many views as necessary to show the invention. The views may be plan, elevation, section, or perspective views. Detail views of portions of elements, on a larger scale if necessary, may also be used. All views of the drawing must be grouped together and arranged on the sheet(s) without wasting space, preferably in an upright position, clearly separated from one another, and must not be included in the sheets containing the specifications, claims, or abstract. Views must not be connected by projection lines and must not contain center lines. Waveforms of electrical signals may be connected by dashed lines to show the relative timing of the waveforms.

> We've already discussed the most common types of views. Don't worry if you don't specifically recognize all of view-types listed – they basically cover every view you can think of that conforms to the rest of the rules presented here.
>
> "All views of the drawing must be grouped together" just means within your application all of your drawings must be located together in a single section with no other part of the application between them.
>
> Regarding the comment about not wasting space – you don't have to overwork this. Make a reasonable effort to fit more than one drawing on a page if it will fit (keeping them in order), but no one's going to go through it and measure to see if you could possibly have saved half a page.

(1) **Exploded views.** Exploded views, with the separated parts embraced by a bracket, to show the relationship or order of assembly of various parts are permissible. When an exploded view is shown in a figure that is on the same sheet as another figure, the exploded view should be placed in brackets.

> This is fairly self-explanatory. Normally I don't think exploded views are better than sectionals – usually you're trying to illustrate how your patent works, not how to take it apart and put it back together.

(2) **Partial views.** When necessary, a view of a large machine or device in its entirety may be broken into partial views on a single sheet, or extended over several sheets if there is no loss in facility of understanding the view. Partial views drawn on separate sheets must always be capable of being linked edge to edge so that no partial view contains parts of another partial view. A smaller scale view should be included showing the whole formed by the partial views and indicating the positions of the parts shown. When a portion of a view is enlarged for magnification purposes, the view and the enlarged view must each be labeled as separate views.

> Favor defining magnified views as shown in the example (Fig. 6-1) over simply breaking a large view into sequential sections distributed across pages whenever possible.

    (i) Where views on two or more sheets form, in effect, a single complete view, the views on the several sheets must be so arranged that the complete figure can be assembled without concealing any part of any of the views appearing on the various sheets.

(ii) A very long view may be divided into several parts placed one above the other on a single sheet. However, the relationship between the different parts must be clear and unambiguous.

(3) **Sectional views.** The plane upon which a sectional view is taken should be indicated on the view from which the section is cut by a broken line. The ends of the broken line should be designated by Arabic or Roman numerals corresponding to the view number of the sectional view, and should have arrows to indicate the direction of sight. Hatching must be used to indicate section portions of an object, and must be made by regularly spaced oblique parallel lines spaced sufficiently apart to enable the lines to be distinguished without difficulty. Hatching should not impede the clear reading of the reference characters and lead lines. If it is not possible to place reference characters outside the hatched area, the hatching may be broken off wherever reference characters are inserted. Hatching must be at a substantial angle to the surrounding axes or principal lines, preferably 45°. A cross section must be set out and drawn to show all of the materials as they are shown in the view from which the cross section was taken. The parts in cross section must show proper material(s) by hatching with regularly spaced parallel oblique strokes, the space between strokes being chosen on the basis of the total area to be hatched. The various parts of a cross section of the same item should be hatched in the same manner and should accurately and graphically indicate the nature of the material(s) that is illustrated in cross section. The hatching of juxtaposed different elements must be angled in a different way. In the case of large areas, hatching may be confined to an edging drawn around the entire inside of the outline of the area to be hatched. Different types of hatching should have different conventional meanings as regards the nature of a material seen in cross section.

This fairly self explanatory rule set is exactly what was described near the beginning of this section.

Arabic numerals are just the regular 1, 2, 3... number characters that you are familiar with. Notice you can actually use Roman numerals if you want, but why would you do this?

You don't need to worry about the last line regarding cross-hatching types matching material conventions; in fact, unless there is a specific reason affecting function, addressing the materials of construction of a specific part is not mandatory, since simple material substitutions to an exiting invention do not meet the novelty requirements to count as a new patentable invention.

DETAILED LISTING OF DRAWING RULES   59

(4) **Alternate position.** A moved position may be shown by a broken line superimposed upon a suitable view if this can be done without crowding; otherwise, a separate view must be used for this purpose.

> It's best to just use another figure. These days, using electronic drawings, this is rather trivial since you can just copy/paste the drawing and move the part.

(5) **Modified forms.** Modified forms of construction must be shown in separate views.

> Modified forms are alternate embodiments (using alternate or including optional features) of your invention. What this basically means it you shouldn't try to show parts that can be interchanged within a single figure. Show a new view that is identical to the baseline embodiment, but with the alternate parts or added features included. With modern drawing programs this is a simple as cut/paste/modify. If an optional feature is simply something that is present or not without affecting the function of other parts, you can just include it in the baseline and call it out as optional in the text.

(i) **Arrangement of views.** One view must not be placed upon another or within the outline of another. All views on the same sheet should stand in the same direction and, if possible, stand so that they can be read with the sheet held in an upright position. If views wider than the width of the sheet are necessary for the clearest illustration of the invention, the sheet may be turned on its side so that the top of the sheet, with the appropriate top margin to be used as the heading space, is on the right-hand side. Words must appear in a horizontal, left-to-right fashion when the page is either upright or turned so that the top becomes the right side, except for graphs utilizing standard scientific convention to denote the axis of abscissas (of X) and the axis of ordinates (of Y).

> As discussed above, the outline of a view is the smallest box that can be drawn around the view encompassing all graphics and text – if you draw these boxes around all images on a page, none of them can intersect.
>
> The "all views on the same sheet should stand in the same direction" rule takes precedence over trying to get as many FIGs on the same sheet as possible.
>
> You may use "portrait" or "landscape" page layouts – not all pages of your drawings have to have the same orientation. On landscapes, note the required orientation.

(j) **Front page view.** The drawing must contain as many views as necessary to show the invention. One of the views should be suitable for inclusion on the front page of the patent application publication and patent as the illustration of the invention. Views must not be connected by projection lines and must not contain center lines. Applicant may suggest a single view (by figure number) for inclusion on the front page of the patent application publication and patent.

> One drawing has to be suitable for the cover sheet of the published patent. Honestly, I can't think of how you could possibly have drawings that adequately describe your invention without one being a good top-level illustration, so compliance should be almost automatic. I also can't think of why one would go to the trouble to suggest a particular view for the cover sheet – the patent reviewer will do his/her job and make that call anyway.

(k) **Scale.** The scale to which a drawing is made must be large enough to show the mechanism without crowding when the drawing is reduced in size to two-thirds in reproduction. Indications such as ``actual size'' or ``scale \1/2\'' on the drawings are not permitted since these lose their meaning with reproduction in a different format.

> Generally you won't actually have to do a 1/3 reduction of your drawings to know if they are too small. Just make sure that all of the lines on your drawings are well spaced with margin. It is normal for some region of your top-level drawing showing the whole invention to be too busy – that's what close-up views are for, and you should use them.

> Scale marks are indeed pointless, and don't really have anything to do with teaching your invention. Sometimes the relative scale of two features might be important, but you have to talk to that in the text. Scale normally has nothing to do with the inventive concepts embodied in an invention. A clock is basically the same idea, whether we're talking about a wristwatch or Big Ben.

(l) **Character of lines, numbers, and letters.** All drawings must be made by a process which will give them satisfactory reproduction characteristics. Every line, number, and letter must be durable, clean, black (except for color drawings), sufficiently dense and dark, and uniformly thick and well defined. The weight of all lines and letters must be heavy enough to permit adequate reproduction. This requirement applies to all lines however fine, to shading, and to lines representing cut surfaces in sectional views. Lines and strokes of different thicknesses may be used in the same drawing where different thicknesses have a different meaning.

> I recommend not going smaller than 0.5 pt lines (which I use for practically everything), except cross-hatch lines can be finer. Remember that you'll never get docked for having lines that are too thick, unless they are simply so thick that they obscure features. The key, as always with drawings, is that they be clear. Look to your example patents for guidance. The lines should be thick enough to appear black when observed on the printed page. If they look at all faint or gray, they are too thin. Also,
> I recommend that if you are creating drawings in another program and pasting them in, make sure you paste them in as an object-based drawing, if at all possible, such as WMF, PNG, PIC.

(m) **Shading.** The use of shading in views is encouraged if it aids in understanding the invention and if it does not reduce legibility. Shading is used to indicate the surface or shape of spherical, cylindrical, and conical elements of an object. Flat parts may also be lightly shaded. Such shading is preferred in the case of parts shown in perspective, but not for cross sections. See paragraph (h)(3) of this section. Spaced lines for shading are preferred. These lines must be thin, as few in number as practicable, and they must contrast with the rest of the drawings. As a substitute for shading, heavy lines on the shade

> Do not be confused; "shading in views is encouraged" does not mean grayscale is preferred. As you read further note that shading using spaced (preferred) or crossed lines is encouraged. Note also that it says "encouraged". If your art skills aren't up to this, you really don't have to do it (after all, in your detailed description you can just make it clear whether that circle on the page is a ball or a disk), but, admittedly, shade lines sure make a ball look much more spherical.

> Also, black shouldn't be used for shadows – black areas indicate color, not shading.

side of objects can be used except where they superimpose on each other or obscure reference characters. Light should come from the upper left corner at an angle of 45°. Surface delineations should preferably be shown by proper shading. Solid black shading areas are not permitted, except when used to represent bar graphs or color.

Commonly very small objects can be shown black in cross-sections, especially if they are so small that the border line makes them almost completely black anyway.

(n) **Symbols.** Graphical drawing symbols may be used for conventional elements when appropriate. The elements for which such symbols and labeled representations are used must be adequately identified in the specification. Known devices should be illustrated by symbols that have a universally recognized conventional meaning and are generally accepted in the art. Other symbols which are not universally recognized may be used, subject to approval by the Office, if they are not likely to be confused with existing conventional symbols, and if they are readily identifiable.

The symbols referred to are those used for circuit diagrams, plumbing diagrams, and the like. Technically, as long as each component is labeled and described, use of a non-standard symbol isn't going to severely weaken your patent, and, with the more relaxed standards regarding enforcement of the drawing rules, a patent examiner will not likely call you on the use of a particular symbol; but you will be smart (and courteous) to use conventional symbols whenever possible. If you're unsure, check out your reference patents, particularly those that predate 1999 (when the USPTO reduced its scrutiny of drawings).

(o) **Legends.** Suitable descriptive legends may be used subject to approval by the Office, or may be required by the examiner where necessary for understanding of the drawing. They should contain as few words as possible.

Legends within a rectangular box separating it from the rest of the FIG are allowable but technically unnecessary, since all components within a schematic must be numbered and described in the DETAILED DESCRIPTION OF THE INVENTION.

(p) **Numbers, letters, and reference characters.**

(1) Reference characters (numerals are preferred), sheet numbers, and view numbers must be plain and legible, and must not be used in association with brackets or inverted commas, or enclosed within outlines, e.g., encircled. They must be oriented in the same direction as the view so as to avoid having to rotate the sheet. Reference characters should be arranged to follow the profile of the object depicted.

The gist of this is that you must use reference numbers, not letters, and don't try to get fancy. Keep it simple, and try to place your reference numbers close to that at which they are pointing.

(2) The English alphabet must be used for letters, except where another alphabet is customarily used, such as the Greek alphabet to indicate angles, wavelengths, and mathematical formulas.

Yes, the U.S. is still an English-speaking country.

(3) Numbers, letters, and reference characters must measure at least .32 cm. (1/8 inch) in height. They should not be placed in the drawing so as to interfere with its comprehension. Therefore, they should not cross or mingle with the lines. They should not be placed upon hatched or shaded surfaces. When necessary, such as indicating a surface or cross section, a reference character may be underlined and a blank space may be left in the hatching or shading where the character occurs so that it appears distinct.

> I discussed this above, but I'll repeat here that I recommend 14 pt font. Reference numbers should be either connected to the part with lines, or on top of the part and underlined. If the number is on top of a part, it must be intersected by no line, including cross-hatching or shade lines.

(4) The same part of an invention appearing in more than one view of the drawing must always be designated by the same reference character, and the same reference character must never be used to designate different parts.

> No number may refer to more than one part, and the same unmodified part must never be identified with two different numbers.

(5) Reference characters not mentioned in the description shall not appear in the drawings. Reference characters mentioned in the description must appear in the drawings.

> Any reference character must exist in both the drawings and the DETAILED DESCRIPTION OF THE INVENTION.

(q) **Lead lines.** Lead lines are those lines between the reference characters and the details referred to. Such lines may be straight or curved and should be as short as possible. They must originate in the immediate proximity of the reference character and extend to the feature indicated. Lead lines must not cross each other. Lead lines are required for each reference character except for those which indicate the surface or cross section on which they are placed. Such a reference character must be underlined to make it clear that a lead line has not been left out by mistake. Lead lines must be executed in the same way as lines in the drawing. See paragraph (l) of this section.

> For more complex inventions, it can be a challenge to pack in all the reference numbers, since their lines cannot cross. Additionally, it's important to make sure that the lead lines don't obscure another feature, and don't get lost amidst cross-hatch lines (so think about what angles your lead lines will be coming from when you do your cross-hatching. Lead lines must not actually touch the reference number from which they emanate.

(r) **Arrows.** Arrows may be used at the ends of lines, provided that their meaning is clear, as follows:
   (1) On a lead line, a freestanding arrow to indicate the entire section towards which it points;
   (2) On a lead line, an arrow touching a line to indicate the surface shown by the line looking along the direction of the arrow; or

> Arrows have specific meanings on lead lines as listed here. The vast majority of lead lines start from near the reference character and terminate at a convenient place inside the border of the indicated part in your drawing, and should have nothing on either end (no arrows, but also no dots, x's, etc).

(3) To show the direction of movement.

> I recommend avoiding the use of arrows to show direction of movement; just talk to it in the text and show the alternate position, either by a second rendition of the part with the appropriate dotted line border (— ·· — ·· —), or, preferably, in a second FIG.

(s) **Copyright or Mask Work Notice.** A copyright or mask work notice may appear in the drawing, but must be placed within the sight of the drawing immediately below the figure representing the copyright or mask work material and be limited to letters having a print size of .32 cm. to .64 cm. (1/8 to 1/4 inches) high. The content of the notice must be limited to only those elements provided for by law. For example, "©1983 John Doe" (17 U.S.C. 401) and "*M* John Doe" (17 U.S.C. 909) would be properly limited and, under current statutes, legally sufficient notices of copyright and mask work, respectively. Inclusion of a copyright or mask work notice will be permitted only if the authorization language set forth in 1.71(e) is included at the beginning (preferably as the first paragraph) of the specification.

> I'm assuming you are doing your own figures, so copyright notices will be a non-issue. If you do have someone else do drawings who want to have a notice, they'll let you know.
>
> A copyright notice for a patent is somewhat superfluous, since legally all compositions are protected by copyright in the United States, whether or not you file for one, and the patent application provides a clear record of your authorship.

(t) **Numbering of sheets of drawings.** The sheets of drawings should be numbered in consecutive Arabic numerals, starting with 1, within the sight as defined in paragraph (g) of this section. These numbers, if present, must be placed in the middle of the top of the sheet, but not in the margin. The numbers can be placed on the right-hand side if the drawing extends too close to the middle of the top edge of the usable surface. The drawing sheet numbering must be clear and larger than the numbers used as reference characters to avoid confusion. The number of each sheet should be shown by two Arabic numerals placed on either side of an oblique line, with the first being the sheet number and the second being the total number of sheets of drawings, with no other marking.

> The "sight" is the entire page minus the margins. Your drawing numbers must be within, not in, the margins, and you must use a larger font than your reference numbers (I recommend 18 pt, which has been formatted in the Application Workbook template. The drawing page number must not infringe upon the drawings, that is, they should not intersect the smallest square you can draw around each drawing that encompasses all graphics and text, including the FIG number (see Fig. 6-2).
>
> The drawing page numbers should look like "1/24" where 1 is the page number and 24 is the total number of drawing pages.

(u) **Numbering of views.**

(1) The different views must be numbered in consecutive Arabic numerals, starting with 1, independent of the numbering of the sheets and, if possible, in the order in which they appear on the drawing sheet(s). Partial views

> Your drawings should definitely be numbered in the order they are addressed in the text, and should appear on the drawing pages in this order. (When would this not be "possible"?)

intended to form one complete view, on one or several sheets, must be identified by the same number followed by a capital letter. View numbers must be preceded by the abbreviation "FIG." Where only a single view is used in an application to illustrate the claimed invention, it must not be numbered and the abbreviation "FIG." must not appear.

> Note that if you have a single drawing (not a single page, which may have more than one drawing, but a single view of the invention), you should not number it. In your text, instead of referring to it as "FIG. 1", just call it "the drawing".

(2) Numbers and letters identifying the views must be simple and clear and must not be used in association with brackets, circles, or inverted commas. The view numbers must be larger than the numbers used for reference characters.

> I recommend 18 pt. font for the FIG labels.

(v) **Security markings.** Authorized security markings may be placed on the drawings provided they are outside the sight, preferably centered in the top margin.

> This pertains to patents that contain government classified material.

(w) **Corrections.** Any corrections on drawings submitted to the Office must be durable and permanent.

> White-Out is now allowed, but please – just fix the image and reprint the page.

(x) **Holes.** No holes should be made by applicant in the drawing sheets.

> Of course, this also implies that you shouldn't use paper with pre-made holes either.

(y) **Types of drawings.** See § 1.152 for design drawings, § 1.165 for plant drawings, and § 1.174 for reissue drawings.

> § 1.152 contains special restrictions on use of dotted lines for design drawings (no dotted lines to show hidden or out-of-plane objects, or alternate positions.

> § 1.165 exempts plant drawings from view numbers and requires color drawings if color is a distinguishing feature of the new type of plant

> § 1.174 states that a patent re-issue application can reuse drawings from the original patent.

Above all else, make sure your drawings are clear and unambiguous.

So now you have the basic rules for your drawings. As you can see, there are quite a few, but don't worry, as I mentioned at the beginning of this section, most of the rules are not currently being strictly enforced. This does not mean that they are unimportant – the spirit of these rules is key to protecting your invention, and that is to be as plain-in-meaning and unambiguous as possible.

Clearly, following the rules diligently is one concrete step in the right direction, but it is easily possible to follow them and still fail to communicate your invention effectively. To check this, there's simply no substitute for a second set of eyes. I'm not talking about a professional. Hopefully that someone you can trust from Chapter 1 is still available to you. It's OK if a previously unindoctrinated reviewer can't make complete sense of your drawings without the explanatory text you will later provide in your specification; but, if your drawings are clear, you should be able to successfully make any reasonably technically astute evaluator fully comprehend your invention using your drawings as the only required visual aids. If you need to stop and sketch to answer any questions, that should be a clue to you that you need another FIG.

> **Your drawings should be such that you can effectively explain your invention using them as your only visual aids.**

6.4 DRAWING QUALITY

I mentioned in the first section of this chapter that enforcement of drawing rules has generally become quite lax. The blunt truth is that to what degree the drawings rules will be enforced is largely up to the specific patent examiner who gets your application as the luck of the draw. (It's actually not totally random – the USPTO tries to facilitate the specialization of patent examiners into specific fields of invention so that they develop thorough working knowledge of the prior art.) Patent examiners' expectations can vary greatly, and having a lax examiner is not necessarily to your benefit. The fact is that you may find that your patent examiner will let you get by with drawings that are genuine garbage...and I've seen some garbage. I don't mean drawings by someone who wasn't a good artist; I mean drawings by someone who didn't even try.

> **Don't rely on your patent examiner to call you on it if your drawings aren't up |to snuff.**

There is more to producing good drawings than simply following the letter of the law. Remember in Section 2.3 when I said that your invention should certainly look quite different than that first napkin sketch by the time it's ripe for your patent? Well, I'll tell you I actually have seen patents where the drawings quite literally appeared to be that original napkin sketch. I mean hand sketches that looked like they took five minutes apiece at best – the "drafter" didn't even bother to pick up a ruler to make straight lines. Technically, that doesn't violate any rule, but I expect that examiner took one look at the drawings and decided he certainly wasn't going to spend more time on them than the applicant. You will find that when people don't view you as taking it seriously, they'll sit back and smile while they watch you place a noose around your own neck.

> **If you don't take your application seriously, don't expect your patent examiner to either.**

PEASE DON'T DO THAT.

I might have to read your patent someday. And besides, when it comes down to it, anything worth doing really is worth doing right. How far do you think that guy's napkin sketches will fly in court if someone ever challenges his date of inventorship? Do you remember that the U.S. gives priority to the first to invent, not the first to patent? So, if some guy shows up with chicken scratch, and another a polished, beautiful, and comprehensive presentation of the invention, who's going to be perceived as the more credible?...and they should be. Lastly, there's simply this – *Never ever put your name on anything that isn't your best work.*

6.5 LET'S GET TO WORK

Never put your name on anything that isn't your best work.

Place your drawings in the Application Workbook as float-over-text images or, preferably, in canvases.

OK – let's move forward. Your next action will be to create your drawings in whatever drawing program you have chosen to use. Once you have completed the line drawings without reference characters and lines, you're going to import (Insert/Picture/From File...) them into the workbook as float-over-text images. Do not import them as inline-with-text images. Inside the program you can readily convert them back and forth, but ultimately you want them all as floating objects. In more recent versions of MS Word, you can place drawing object assemblies on canvases, and that's useful, because it will stop the screen from jumping over to the cursor (when you're zoomed in) every time you click the background (one of the lamest and most difficult to explain features of MS Word).

Hold off on reference lines and characters in your figures until after pasting them into the Application Workbook.

Create only the line drawings for the FIGs in your drawing program, but not the reference characters and lines (e.g. all FIG labels, reference numbers and connecting lines, and reference planes and outlines denoting views provided by other FIGs shown in Fig. 6-2). The reason is that all reference characters and associated lines are going to be inserted as text boxes and graphics floating over your images in MS Word so that you will be able to dynamically link them to your specification text (this will save you hours of tedious corrections – more on this later when we discuss drafting the detailed description of the invention). It's OK for you to just make draft drawings at this time if you feel you must, but I do recommend that you do your best to make them as finished as possible from the outset. There's really no advantage to putting it off.

Once you have pasted your FIGs in the Application Workbook, complete the reference lines only, but not the element numbers, since you do not yet know what they will be.

Once you have completed final drafts of the drawings, you should import them into the drawing pages in the Application Workbook in the order you intend to address them in the text. Place the FIG numbers next to them, and reference lines showing relationships to other views, but still hold off on element numbers for now – you won't know what numbers to assign to which elements until you write your DETAILED DESCRIPTION OF THE INVENTION in Chapter 8. Making the reference lines a live part of the MS Word document will mean you can edit them later if you need to, say, add in an extra FIG and some of the other FIG numbers need to shift. You can overlay text boxes containing placeholder text such as "???" and lines connecting them to the elements of the drawing now, and leave only the task of replacing the question marks with links to the reference numbers in the text for later, if you prefer.

> **Action Step 6**
>
> Create your drawings. When complete, import them in the order they will be referenced in the text into the drawing pages of Section 3 of the Application Workbook. (If they are drafts, don't take the time to paste them in. Come back to this step when you complete them later.)
>
> Once in the Application Workbook, overlay your reference lines and FIG labels.
>
> Hold-off on reference numbering, but you may put in placeholders if you wish.

The Wizard of Menlo Park

Some people have buildings, roads, or even cities named after them. Fewer leave their names to theorems and mathematical equations, fundamental dimensionless scaling parameters, or universal constants. But how many become a word in the English language?

Thomas Edison trained as a telegraph operator at age nineteen and, not coincidentally, his first profitable patent was the quadruplex telegraph, which doubled the number of signals a single wire could carry from two to four. In 1876 Edison used the profits from his early success to establish the world's first industrial research facility in Menlo Park, New Jersey and chalked up some 1,093 U.S. Patents over the next 64 years – on average about one every three weeks (apparently most anything developed at the company). While, like the quadruplex, most of these inventions were improvements to existing technologies, his unprecedented introduction of the seemingly magical phonograph to a world where sound had never before been recorded catapulted the him to great fame. Edison was also an astute and ruthless businessman, founding Edison Electric (later GE) which, during the "War of Currents", campaigned to vilify AC power distribution (vs. his DC system) by performing public animal electrocutions. Amazingly, Edison achieved all that he did with the benefit of only three months formal education, for which reason his approach to invention was extremely empirical. In his successful efforts to develop the first practical incandescent light bulb filament, Edison famously tested more than 3000 different material candidates. Consequently, to this day engineers not uncommonly refer to development approaches that rely heavily on experimentation as "Edisonian". When Edison said "Invention is 1% inspiration and 99% perspiration," he meant it.

7 Starting the Specification

7.1 LET'S START WRITING

The upfront sections of a patent make the case for the usefulness of the invention.

In this chapter, we will complete a laundry list of the introductory portions of the specification. These sections are not fluff – in particular, they are the place where you make a case for the usefulness (and to some extent non-obviousness) of your invention, and you will find that working them first will really help organize your thoughts.

7.1.1 Title

Spend several minutes thinking about the message your invention title conveys.

OK, have you at least draft drawings completed? Let's get down to this writing business. Step 1 is to select a title for your invention. When Shakespeare's Juliet so famously asked "What's in a name?", she (or rather, Shakespeare) meant the answer to be nothing, but, then again, Shakespeare didn't write patents. In our earlier example, a patent entitled "LAWN MOWER POWERED BY ELECTRIC MOTOR" may come across as obvious, whereas one bearing the title "REDUCED NOISE LAWN MOWER" may not. So, while the selection of a title for your patent isn't worth hours of thought, do spend a few minutes thinking about what you want the title to convey.

> **In particular, remember that the title of your patent:**
> - Is the first thing the patent examiner will see.
> - Will be the first thing seen by others when they do patent searches, and the deciding factor of whether or not they click through to look at the first page of your patent.

Your choice of title will certainly affect how easily patent database searches find your patent.

From a legal standpoint, the title does not make much difference and does not have to be original (two patents can have the same title), but should generally describe the invention. Nonetheless, you may make decisions about word choices based on how easy you wish your patent to be to find in general searches. Remember, you're applying to receive special rights in exchange for teaching your invention to the world, so, strictly speaking, to deliberately use language that makes the patent more obscure violates the spirit of the process. Nevertheless, sometimes inventors feel that it is in their best interest to keep their patent quiet until they are ready to field a product (at which time the patent number is generally noticeably marked on the product and anyone can find it easily). Since patent applications take about two years to be published, often this is a non-issue, but sometimes patent products can take significantly longer to develop into a marketable device.

Your title should be short and to the point.

I would be remiss if I did not also convey the USPTO's direction that your title be as short and specific as possible. In spite of this, technically your title may contain up to 500 characters. If you want to be smart, keep it short and sweet; if you want to be a smart (beep), you can really get your patent examiner's goat right out of the gate by coming up with a nice paragraph-long title. For reference, this paragraph is exactly 500 characters long (counting spaces) if I stop right here: X.

By the way, the USPTO requests (but does not require) that paragraphs within the specification be numbered [0001], [0002], [0003], etc. We aim to please, so you'll find this format already embedded in the Application Workbook. Your patent title should look like this:

[0001]   REDUCED-NOISE LAWNMOWER

Action Step 7

Go to Paragraph [0001] on the specification pages of Section 3 of the Application Workbook and type your patent title where indicated. Use all capitals, 12 point Times New Roman font or equivalent.

Remember, you can always, and may well, change it later after you've completed more of the specification text.

7.1.2 List of Inventors

Leaving out an inventor, or adding an extra one in, can invalidate a patent.

The second part of the title section is a list of the inventors, which must include the address and citizenship of each (you do not need to be a U.S. citizen to get a U.S. patent). Reporting citizenship is optional, but I recommend including it. The only catch here (and on the filing forms we'll cover later) is that you must include only and all actual inventors. Omitting a legitimate inventor can invalidate a patent. Listing extra names who are not actual contributors to what is specifically claimed, even just to be nice, can also render the patent invalid. You may be interested to learn that of the 1093 patents claimed by Thomas Edison, only a small fraction are for things he actually invented. The bulk of them actually describe inventions of his employees, who aren't even acknowledged. Apparently one could get away with a lot at the turn of the century. But that won't fly now (well, it might if you were Thomas Edison), so stick to the facts.

Format your list of inventors as follows:

[0002] Inventor(s):   Thomas A. Edison
[0003]   37 Christie Street
[0004]   Menlo Park, NJ 08820
[0005]   Citizenship: United States

[0006]   Nicola. Tesla
[0007]   37 Christie Street
[0008]   Menlo Park, NJ 08820
[0009]   Citizenship: United States

Action Step 8 — Underneath your title in the Application Workbook, replace the inventor information with yours, adding and deleting from the list as required.

7.2 BACKGROUND OF THE INVENTION

Focus on identifying problems in the known art that your invention resolves.

Following the title and inventors list comes a section called, and which you should entitle, the "BACKGROUND OF THE INVENTION", comprised of two subsections. The first, the "Field of the Invention", is very short, normally single sentence stating the general family of products or methods of which your invention would be considered a member. The second part, the "Description of the Prior Art" speaks directly to the requirement that your invention be useful to be eligible to receive a patent. You need not overwork this section – if you take a quick tour of the USPTO's database just typing relatively recent patent numbers in at random, you will quickly realize that you don't have to worry much about objections to your application on the basis of usefulness. More importantly, however, a comprehensive listing of all of the problems that your invention intends to solve, known as the "objects of the invention", will strengthen the interpretation of your claims, and that should be your primary focus.

7.2.1 Field of the Invention

Your example patents will provide you a quick idea as to what to designate as the Field of the Invention.

Every patent contains a statement describing the specific field or technology area to which it pertains, which serves as one of the searchable fields that help people researching the prior art to find that for which they are looking. There is no rule regarding specific wording; all that is required is that the statement must generally accurately describe the general category of inventions into which your patent falls. Here are some examples I grabbed at random:

"The present invention relates to concrete construction apparatus and, more particularly, to apparatus for forming a concrete deck."   U.S. Pat. No. 4,342,440

"This invention relates to semiconductors and more particularly to MOSFET memory devices."   U.S. Pat. No. 543,210,6

"The present invention relates to polarimetry, ellipsometry, reflectometry, spectrophotometry and the like, and more particularly to systems for, and methods of controlling radial energy density profiles in, and/or cross-sectional dimensioning of electromagnetic beams."   U.S. Pat. No. 6,590,655

Remember you're describing the field of the invention, not the invention itself.

Anyway, you get the gist. Actually, as I searched for examples I was reminded of how many patent writers don't seem to get it – I found far more bad examples than good ones as I just typed random patent numbers in the USPTO's database search engine. Remember, you're describing the field of the invention, not the invention.

Action Step 9

In the BACKGROUND OF THE INVENTION section of the Application Workbook, describe the field of your invention under the subheading "Field of the Invention".

Finish the sentence: "The present invention relates to…" or "This invention relates to…"

The beginning part of your BACKGROUND OF THE INVENTION should look like this:

[0010]   BACKGROUND OF THE INVENTION

[0011]   Field of the Invention

[0012]   The present invention relates to…

7.2.2 Description of the Prior Art

Traditional writing style rules count for very little in the patent world.

OK, moving along we're now into sections that actually are going to require a little creative writing. Not your favorite subject back when you were in school? Not to worry – I actually want you to be generally as uncreative is as possible. Your writing should be direct and straight to the point. Forget all of the basic style rules your English teachers evangelized (if you're a typical engineer, you've long forgotten them already, if you ever learned them at all). You won't get any extra points for nice prose here.

One very specific style point that I want to emphasize that runs 180 degrees to what you learned in school relates directly to vocabulary. In creative writing, it is bad form to unnecessarily reuse the same word repeatedly, but rather to exercise the available vocabulary by using different words or expressions to refer to the same concept or object to make the reading more interesting. In your patent, you should do just the opposite. To

Refer to the same thing with the same words every time.

the greatest extent possible you should refer to a specific subject in exactly the same wording every time. Always remember that you are trying to minimize ambiguity. Unlike creative writing, where you try to engage the reader by letting them participate, here your goal is to be as closed to alternate interpretations as possible. Create specific names for the components of your invention, and stick with them. There is also a very practical advantage to this. We're getting a little ahead of ourselves, but I'll just forewarn you that every element of your invention described in the detailed description will need to be identified with a number, and I can tell you that you will find it a royal pain in the you-know-what if you try to enter all of the numbers as you type the text. What you will do is go back and enter the numbers after you've composed the text using a keyword search – and that's going to be a lot easier if you've used the same wording in every instance.

Use lists to avoid wasteful transition language.

We're also going to use lists to the greatest extent possible to minimize the effort required to work graceful transitions in your prose. You don't need them. Again, you're not writing a the next great work of fiction (hopefully). Always remember the purpose of the document you are creating is to teach an invention that is novel, useful, and non-obvious,; hence in each section of this workbook, always ask yourself to which of these attributes that section speaks.

General Introduction to Field of Invention

Your introductory paragraph should address, in general terms, for what your invention is useful.

Your first real paragraph will basically take the general form of an introduction, but it has one specific point – only here will you address in broad terms the usefulness of your invention, and recall that usefulness is one of the three cardinal requirements to receive a patent. In this first paragraph or set of paragraphs you will set the stage by presenting the general need or purpose that your invention serves. Don't overwork this! In most cases a single paragraph will do. As you did your patent search, you may have noticed that some patents have almost nothing in this section at all, which should convey to you that requirements here are pretty lax. Nevertheless, I do recommend that you try to put some quality into your specification. Conversely, you may find you are enjoying writing about the topic, and you are free to go on at some length. Keep in mind, however, that your specific objective is to set the stage for your invention, and note in particular you are winding towards the next segment where you will list the deficiencies of inventions created for the same purpose that are known to be in the art that your invention remedies. In this paragraph, you should make reference to related inventions, particularly those that serve as examples that broadly outline different schools of approach to the want or need addressed by your invention, but, again, don't overwork this. Naturally, if your invention is an improvement to something already in the art, that invention should be referenced.

Find a blank page daunting? The first paragraph is almost always the most difficult, but, fortunately, when you performed your patent search you earmarked two to three patents you liked (you did, didn't you?). Now is the

> **Your introduction will say much of the same things as those of your example patents, but you must use your own words.**

time to print them out. There's no reason to re-invent the wheel – draw from them. Do not outright plagiarize them, however. In the U.S., all written works are copyrighted to the author by default, whether or not the author formally files for copyright protection. One way to focus your thoughts is to keep in mind what the next paragraph will say. What you are looking for in your introduction is to create a minimal context into which that next paragraph will fit. In this case, I can foreshadow that you will want the next paragraph to go straight to listing the problems known to be in the art that your patent is going to resolve, and that is the stage you are trying to set.

Flow logic for a typical introduction:

1. Make a statement about something people do, or want to do;

2. Follow with how they currently do it, or why they can't do it;

3. Finish by describing by what means (methods and/or tools) they currently do it, or what they do instead.

Action Step 10: Immediately beneath the "Field of the Invention" subsection in the BACKGROUND OF THE INVENTION section of the Application Workbook, write your introduction under the subheading "Description of the Prior Art"

Problems with Known Art

> **Only list problems in the known art that your invention lessens or resolves.**

With your introduction completed, you are now onto the key step of identifying known problems that exist in the art that your invention solves. Take a look at your example patents. Most have a number of benefits, and I expect you can think of more than one improvement your invention brings to the table (but technically you only need one). Effectively, here you are listing the benefits of your invention, but stating them in the form of shortcomings to other inventions already out there. You can do this in either one of two ways. You can either (1) write a short paragraph describing the deficiencies in the known art that your invention improves upon, or (2) you can just list them in complete sentences. It will be easier for everyone (including the examiner) if you simply list them like this:

[0015] One problem with lawnmowers known to be in the art is that they are loud, often requiring the user employ hearing protection and inflicting discomfort upon nearby bystanders.

[0016] Another problem with lawnmowers known to be in the art is that they require gasoline, which is often inconvenient to obtain and store. Additionally, the required storage of gasoline presents a fire hazard.

[0017] Another problem with lawnmowers known to be in the art is that the gasoline-powered motors, which are often started by pulling a spring-loaded cord that turns the motor crank, are difficult to start, particularly when not having been recently started or in cold weather.

[0018] Another problem with lawnmowers known to be in the art is...

Do you see how we're setting the stage to introduce our electric lawnmower? And, that's exactly what we will do in the next section.

The problems your invention resolves don't have to be unique, just the way it resolves them.

The benefits of your invention do not have to be entirely unique, and, as a matter of fact, none of them need be new at all. The fact that other inventors have found alternate ways to solve the same problem(s) is OK. For your invention to be novel only requires that your solution be unique, not the problems addressed. For instance, if you've invented a swivel for use on shopping carts so that the wheels never get stuck or chatter (if only), you could list stuckage, and chatter as two problems with the known art. Someone else could propose an invention that solves the same problem by inventing a hover-cart, and their specification would naturally list the same two problems with the known prior art.

Action Step 11

Write a paragraph or list describing problems with the known art that your invention remedies underneath your introduction in the BACKGROUND OF THE INVENTION section of the Application Workbook.

Even if you choose to use a list format, remember to always use complete sentences.

7.3 Brief Summary of the Invention

At long last you are finally going to start writing about your own invention. Again, you don't need to be wordy at all. Generally the content of this section usually entails to varying degrees one or more of the following:

1. A terse summary of the substance or general idea of the invention.

2. The advantages of the invention and how it solves the problems you identified in the BACKGROUND OF THE INVENTION.

3. Statements of the object of the invention.

You will notice an enormous variety as you review the example patents you printed out, which should tell you that you have a lot of flexibility here. Generally, for simplicity I like to start out by addressing the ways that the invention helps with or fixes the problems identified in the BACKGROUND OF THE INVENTION. Start with a statement something like this (or exactly like this one if you wish):

[0021] BRIEF SUMMARY OF THE INVENTION

[0022] While some lawn mowers known to be in the art circumvent some of the above listed problems, all of these and other problems are mitigated or eliminated by the lawnmower of the present invention.

Follow a brief synopsis of the invention with a list of ways your invention resolves the problems identified in the BACKGROUND OF THE INVENTION.

From here you should follow with a very terse summary of the substance/general idea of your invention, after which you may just list one-by-one (in direct correspondence to the list you made in the BACKGROUND OF THE INVENTION) the ways by which your invention solves or reduces those problems. You can make brief descriptions of the specific means by which these improvements are accomplished, referring back to the terse summary in the first paragraph.

Always refer to your invention as "the [whatever] of the present invention".

Notice that I referred to the lawnmower in question as "the lawnmower of the present invention". You should always refer to your invention within your specification in kind, and it's best if this name corresponds to the title of your invention. Call it "the ditch digger of the present invention" or "the neural serial jack of the present invention" or "the toenail extractor of the present invention" or whatever, but pick something that begins with "the" and ends with "of the present invention", and stick with it.

Statements of the object of the invention usually take a form substantially equivalent to the following:

[0024] An object of the present invention is to provide a reduced noise lawnmower whereby, the need for the user employ hearing protection is eliminated and discomfort to nearby bystanders is reduced.

[0025] Another object of the present invention is to provide a lawnmower that does not require gasoline, that is often inconvenient to obtain and store, and for which the storage thereof presents a fire hazard.

[0026] Another object of the present invention is to provide a lawnmower that is easier to start than gasoline-powered lawnmowers, which are often started by pulling a spring-loaded cord that turns the motor crank, and are therefore difficult to start, particularly when not having been recently started or in cold weather.

[0027] Another object of the present invention is...

To avoid redundancy, most patents will contain an explicit list of either problems known to exist in the prior art or objects of the invention, but not both.

Note that this object list substantially addresses the same issues that we included in the background of the invention. Generally, you will not see both listed quite so explicitly. A list of statements of the object of the invention will often follow a background section that provides a less specific introduction of the known issues. I recommend that you will find it easier to organize your thoughts and get the content onto the page if you start with listing the known problems in the art (in the BACKGROUND OF THE INVENTION), and then just listing how your invention addresses these problems in the BRIEF SUMMARY OF THE INVENTION, foregoing an explicit list of objects of the invention here. So doing will focus this section on the key physical characteristics and functionality of your invention. There is an exception, of course, if your invention does not address an existing problem or need, but rather creates a whole new capability, in which case the flavor of your narrative will be about how your invention allows users to do something they previously couldn't. This will be unusual, however. Almost all inventions replace something else or address some existing issue.

> **Action Step 12**
>
> Write your BRIEF SUMMARY OF THE INVENTION under the heading provided in the Application Workbook.
>
> Start with the provided introductory line if it suits you, or change it at your option.

7.4 BRIEF DESCRIPTION OF THE SEVERAL VIEWS OF THE DRAWING

All FIGs must appear in the BREIF DESCRIPTION OF THE SEVERAL VIEWS OF THE DRAWING.

Since you've already completed at least draft drawings, this section should be a piece of cake. Here you are going to list your drawings in order (and your drawings should be numbered in the order they are introduced in the text of your DETAILED DESCRIPTION OF THE INVENTION (the next section). If, as you write, you end up changing the order of the drawings, you must also come back and change the order here.

Each list entry should be a single complete sentence in the following form:

> [0000]   FIG. X is a(n) (sectional/exploded) view from the [view angle] (of the [subpart or section]) of a [invention name] made according to the present invention (where/with...)(,shown to advantage).

Each FIG should be identified as an ordinary (but you don't have to write "ordinary"), sectional, or exploded view.

The square brackets indicate mandatory places you will need to fill in information, and the regular parentheses are optional words or clauses. So, each line will always start out either "FIG. X is a view from the...", or "FIG. X is a sectional view from the...", or "FIG. X is a exploded view from the..." (where X is the figure number). These are the three basic types of views:

Ordinary view – If you just say view, you're indicating a drawing of the invention or component of the invention as it would appear when observed by ordinary means.

Sectional View – A view where part of the invention has been cut away to show internal components/features.

Exploded View – An exploded view shows the invention disassembled, to illustrate how the components fit together.

After you introduce the view type, the next clause will specify either the invention or a subpart of the invention, always finishing with "made according to the present invention". From here, your sentence may simply end, or continue to specify something distinct about this particular illustration or embodiment that distinguishes it from other drawings, say the particular position of a movable component or that an extra feature had been added. Finally, if your view represents an enlargement of a portion of a different drawing, you may finish with the phrase "shown to advantage", which means exactly that. This is easy – it will be very clear to you if you peruse a few examples:

Each description should finish by stating what's unique about that particular FIG.

[0031]   BREIF DESCRIPTION OF THE SEVERAL VIEWS OF THE DRAWING

[0032]   FIG. 1 is a view from the side of a lawnmower made according to the present invention.

[0033]   FIG. 2 is a view from the top of a lawnmower made according to the present invention.

[0034]   FIG. 3 is a sectional view from the side of a lawnmower made according to the present invention.

[0035]   FIG. 4 is a sectional view from the side of a lawnmower made according to the present invention with internal components removed to show internal cavities and passages.

[0036]   FIG. 5 is a sectional view from the side of the region in the vicinity of the motor and motor mount of a lawnmower made according to the present invention with vibration isolators incorporated into the motor mounting pylons, shown to advantage.

Before you start, here's a tip that will help you if you have a lot of drawings: You can designate the number in each FIG description as a bookmark. (Highlight the number, select Insert/Bookmark..., and put a descriptive title of the figure in the list.) Then, whenever in your text you need to refer to the drawing, and when you label your drawings, you can just insert a reference to that bookmark. Alternatively, if you want a little bit more of a challenge, you can format your FIG descriptions as a custom formatted number list, and accomplish the same thing (but you'll need to embed some cross-references in the list if you have some drawings with A, B, C,... variations). I don't recommend you insert the figure numbers as

If you have lots of figures, use the cross-reference type of your choice to avoid having to manually update figure references throughout the document in the event of a change.

captions – if you do end up with any A, B, C, etc. versions of the FIGs they won't automatically number, and, since the figure references in the BRIEF DESCRIPTION OF THE SEVERAL VIEWS OF THE DRAWING contain text but the figure captions don't, an automatically generated figure table won't work.

However you opt to do it, this way, if 90% of the way through writing your patent you decide you need to add an extra figure somewhere in the middle, all of your existing references can easily be automatically updated just by editing the FIG list in your BRIEF DESCRIPTION OF THE SEVERAL VIEWS OF THE DRAWING. What you don't want to do is use dead figure references throughout the text to twenty different drawings, and then later realize you need to insert a new one right ahead of FIG. 1, requiring you to manually renumber every FIG and FIG reference in the specification. But, of course, the choice is yours.

Don't expect to not find a need to change FIG numbers at some point – you probably will.

Plan accordingly.

Action Step 13 — Write your BRIEF DESCRIPTION OF THE SEVERAL VIEWS OF THE DRAWING under the heading provided in the Application Workbook.

Detailed Description of the Invention

8.1 Detailed Description of the Invention

The DETAILED DESCRIPTION OF THE INVENTION doesn't specify what counts as your intellectual property.

Now comes the real work. In this section you will provide a detailed description of example embodiments of your invention that illustrate each claim...*example* embodiments. This section does not, and I repeat, *does not* legally define what intellectual property does and does not belong to you. That is accomplished by your CLAIMS, which are addressed in the next and the most important chapter with respect to protecting your interests. Recall that your patent teaches others your invention. Every part of the specification you are writing primarily exists for that intent. You get to make claim to intellectual property in the CLAIMS section in exchange (and as an incentive) for doing this, and that's the section that benefits you. Every other part of your application is for the benefit of others (it is, after all, better to give than to receive), and to that end, this section is the most important.

Traditionally, the example embodiment is about proving usefulness.

An invention must actually work to be useful.

Nevertheless, as I have previously pointed out, a secondary purpose for which the remaining sections of the specification can be used is to make a case proving the originality, non-obviousness, and usefulness of your invention. This section won't really do all that much regarding the first two, but it must address usefulness in that it fulfills a requirement for your application to show a credible example of how your invention could be reduced to practice. The argument basically goes that if you can't teach someone how to actually build it, then you haven't invented it.

This is not to say that the detailed description is irrelevant to your CLAIMS. This section will serve as the primary basis of interpretation of your CLAIMS, and no CLAIM can be made that is not addressed in the DETAILED DESCRIPTION OF THE INVENTION. For instance, for our electric lawnmower I cannot claim a special propeller design if I have not provided a specific description of the novel features of the propeller. Take a look at the CLAIMS in your example patents. Clearly they are terse and do not stand on their own. The DETAILED DESCRIPTION OF THE INVENTION partially defines the invention from a technical standpoint, leaving the CLAIMS to specify within the technical content what's specifically yours and what's generic.

The DETAILED DESCRIPTION OF THE INVENTION does serve as a basis for interpretation of the CLAIMS.

8.2 PART NUMBERING

Every part name mentioned in your text must be associated with a unique number.

Every element of your invention to which you refer in your patent specification will need to be identified by a unique number, which will correspond to labels in your drawings. The numbers should be straight inline with the text (not superscripted), following the part name in each instance it occurs, separated by a single space. They should be of the same font and character size.

Part Numbering Ground Rules:

- Generally, all parts in your drawings should be labeled with a number. You should avoid extra stuff in your figures you don't intent to mention in the text.

- There may be no labeled parts in your drawings that are not mentioned in the text, i.e., there should be no number in the drawings that is not in the text.

- The same part must be labeled with the same number in all drawings and anywhere it is used in the text.

- Numbers may not be recycled. You cannot redefine a number later in the text. New parts must have their own numbers.

- The parts must be numbered in the order they are mentioned in the text. They do not have to be in the order they appear in the drawings.

- Numbers may refer to parts, but also features. For instance, I may identify a bar 1, perforated by a hole 2 near the center, and a hole 3 close to one end. Generally, it's a good idea to use a number for every feature you talk about in the text. Admittedly, sometimes it's a judgment call whether something needs to be numbered. For instance, where, hypothetically, I have already defined a "bolt 9" and defined what constitutes forward and rearward in my drawing, I can at a later point refer to "a hole 10 near the front tip of the bolt 9'" or alternatively "a hole 10 near the front tip 11 of the bolt 9". In the second case, I have assigned a number to the "front tip", which will require me to then label this region of the bolt 10 in the drawings. In general, in such a case as this, as long as I have forward and rearward clearly defined (more on this later), I would probably not opt to increase clutter in my drawings with the extra number.

  On the other hand, occasionally an examiner will take issue if no number is associated with an important region or feature, but I've never heard of an objection to the addition of an extra number added to clarify a descriptive reference, so, when in doubt, just put in the extra number.

- You should also label all pronouns that refer to numbered items with the number of the part or feature to which they refer. Personally, while the number labels preclude any debate regarding a pronoun's antecedent, I recommend, for practical reasons, minimizing the use of pronouns in the specification.

- Always refer to a single part by the same name.

Treating numbers as part of the element name reduces redundant descriptors.

These numbers serve the very specific purpose of eliminating all ambiguity to what the text refers. They also make writing somewhat quicker, actually. If I want to refer to a hole at the front end of a widget, without the numbers, I'd have to write "the hole at the front end of the widget" at every instance. Because it's numbered, I can simply describe a "hole 32 located at the front end of the widget" when I introduce it the first time, and thereafter just write "hole 32".

8.3 VOCABULARY

Let your example patents teach you useful jargon.

Use your example patents to identify particular vocabulary that is specific to your field of invention. Generally, there are no compulsory rules about terminology, but if you see a term in an example patent that clearly conveys special meaning, and, in particular, a definition of which you are not familiar, take the required steps to understand it. I most cases, if the context does not make the meaning of the vocabulary obvious, a dictionary will have the meaning for which you are looking. Use the customary jargon – it'll help you be clear, and can save you the trouble of lengthy descriptions.

Detailed Description of the Invention

"It is to be appreciated that" statements (ITBATs) go with patents like butter with bread.

Lastly on vocabulary, there is a single handy dandy phrase that I want to bring to your attention. You will see this phrase used over and over in the patents you review, and you should feel free to do the same, and that is the "It is to be appreciated that..." statement, A.K.A. "It is to be understood that...". This is a very general way of leading into a sentence about just about anything. This common buzz phrase can save you the trouble of finding more creative ways to introduce ideas and general C.Y.A.-ish (that's cover your...) language, and can also interject, anywhere, a side statement regarding sundry variations on a theme, equivalent substitutions, etc.

Whatever should be said in a specification, but isn't is a missed opportunity – ITBATs expeditiously tie up loose ends.

As always, check your example patents. I can't imagine writing a patent where I didn't introduce a number of things I just wanted to stuff in there somewhere with "It is to be appreciated that..." When you have finished your specification, during your proofing process you should specifically go through it with the single task of asking yourself "are there any "It is to be appreciated that..." statements, or "ITBATs" as we'll henceforth call them, that you want to add. They can even be added as one-sentence paragraphs if they don't belong anywhere else, or even in a list. Always remember nice flowing prose doesn't amount to a hill of beans in a patent specification. Get the information into the text, and put in as many ITBATs as are expedient to that objective.

One excellent use of ITBATs is to anticipate and head-off obvious changes to your invention.

The absolute best use of the ITBAT is to anticipate and head off opportunists who, hypothetically realizing that you're onto something big, would try to file patents for using your invention with simple modifications or in conjunction with other parts. "Wouldn't they be violating the non-obviousness rule", you ask? (If you asked that – good! you're learning...but, remember what I said before about non-obviousness. These days, the USPTO does not make it its business to get too involved in deciding whether or not an invention, or a change to someone else's invention, is obvious.) Such lame patents (believe me, they are common) can often be ultimately defeated, but that's very expensive, and frankly, if it comes down to litigation, you've already lost (and the lawyers have already won), and wouldn't it be easier to just cut them off before they get started? This very handy form of the ITBAT, which you should not do without, comes in the general form (or variations on the theme):

> "It is to be appreciated that [describe change or substitution to the [invention name] 'of the present invention'] without altering the inventive concepts and principles embodied therein".

Let's clarify with a few examples:

> [0043]   It is to be appreciated that there are numerous obvious methods by which a means of recharging may be incorporated into the robotic law enforcement unit of the present invention without altering the inventive concepts and principles embodied therein.
>
> [0028]   It is to be appreciated that the holographic pet of the present invention may be made to resemble any existing or imaginary animal without altering the inventive concepts and principles embodied therein.
>
> [0032]   It is to be appreciated that the operating characteristics of the marshmallow compressor of the present invention may be altered by the addition of supplementary cavities, either within the housing or attachments made to the housing, contiguous in any place with any of the internal passages of the apparatus without altering the inventive concepts and principles embodied therein.
>
> …but it is to be appreciated that any orientation of the fuel feed passage 12, either within the housing 1 or an attachment made to the housing 1 of the deuterium engine of the present invention will not alter the inventive concepts and principles embodied therein.
>
> [0016]   It is to be appreciated that the fabric of the straps 13, 14 of the undergarment of the present invention may be freely substituted without altering the inventive concepts and principles embodied therein.

8.4 Let's Write

Consider starting by defining terms relating to directions and/or orientation.

Depending on the extent and complexity of your invention, the DETAILED DESCRIPTION OF THE INVENTION can be quite long; but, by taking a methodical approach you can keep the task tractable. To get you started, may I suggest a general statement defining any directional/orientational references you plan to use such as "top" and "bottom", "front" and "rear", etc. Something like this:

> [0039] An embodiment to be preferred of the [YOUR INVENTION NAME] of the present invention is here and in figures disclosed. For clarity, within this document all reference to the top and bottom of the [YOUR INVENTION NAME] will correspond to the [YOUR INVENTION NAME] as oriented in FIG. 1, the top of the figure when oriented such that the text is upright corresponding to top of the [YOUR INVENTION NAME], and the bottom of the figure when oriented such that the text is upright corresponding to the bottom of the [YOUR INVENTION NAME]. Likewise, all reference to the front of the [invention name as it appears in title] will correspond to the leftmost part of the [YOUR INVENTION NAME] as viewed in FIG. 1 when oriented with the text upright, and all reference to the rear of the [YOUR INVENTION NAME] will correspond to the rightmost part of the [YOUR INVENTION NAME] as viewed in FIG. 1 when oriented with the text upright.

Clearly I could just have well structured this as a list instead of using a paragraph format.

Directional definitions eliminate ambiguity and can save many words throughout the remainder of the specification.

Generally, you won't see this in most patents, but it can be a great convenience in that now words like top, bottom, front, back, up, down, forward, rearward, etc. have unambiguous definitions within the context of the specification, so can now be used freely. To make the example shown above work, obviously a FIG. 1 showing a side view of the invention would be required. You can also add a sentence defining left and right, if it will be useful to you as you refer to your figures. Feel free to copy exactly this example and modify as required to meet your specific needs.

Do not try to put in reference numbers while you compose the text.

With the coordinates established, you are readily to dig in and write your DETAILED DESCRIPTION OF THE INVENTION. If you try to add in the reference numbers at the same time you compose the prose, you'll go crazy, so don't do that. There's a powerful technique we'll come to later using MS Word's search tool and dynamic cross-references that will make it much simpler to just come back and add the reference numbers after you've completed the text.

Here is a step-by-step way to organize your efforts:

1 You may want to start by making an outline, but realize that you effectively did that when you created your drawings. The specification basically describes those drawings and explains how the depicted devices function; so it will follow them closely and making an outline essentially amounts to figuring out a logical order to address them, one-by-one. You must reference each FIG at least once in the text. You will, therefore, want to have them next to you as you type. Print them out and sort them into the order you intend to talk about them. They may be rough drafts at this point, but they should be complete enough to have all of the components you plan to show in each view.

2 Now, start with the first set of views depicting the most basic embodiment of your invention. Going through those views, you are going to create a list-like set of statements or paragraphs that introduce the parts and features (e.g. holes, passages, grooves, tabs, grips, etc.) one-by-one and describe their basic inter-relationships. You may start off with a statement like this:

> [0044] Referring to the figures, the [YOUR INVENTION NAME] of the present invention comprises, generally:

You are not at this stage going to describe how your invention works, just introduce the components in terms of their shape and relative locations, whether they are free to slide relative to one another, etc. If this is your first patent, I recommend you start by reading through this part of the two to three example patents you printed out to get the gist and style hints. Try to move generally from one end to the other, rather than jumping around. Each time you mention a part, mark it on your printouts with a highlighter (or put an "X" on it, or whatever). When all of the parts on this set of drawings are checked off, this step is complete and you can move on. Again, do not try to put the part labels (the drawing reference numbers) in at this time. We'll come to that later. For now, just try to use the same name for each part each time you refer to it.

3 Now that you have all of the parts defined, write one or more paragraphs describing the function of your invention. If you are referring to additional views that show parts in different positions, make sure you refer to each figure specifically. The correct form of a figure reference in the text is "FIG. X" where X is the figure number. You may use a letter in the figure name (e.g. "FIG. 3A") to group but distinguish between views that bear a relationship, such as two views that are otherwise identical that simply show a moving part in two different positions. If you bookmarked the FIG numbers in your BREIF DESCRIPTION OF THE SEVERAL VIEWS OF THE DRAWING, or decided to use live captions in the FIG labels, go ahead and put the FIG callouts in as cross-references at this time. If so doing tends to wreck your train of thought, just put in placeholders like "FIG. ???" and come back to plug in the cross-references.

As you work through the text, you may discover that you wish you had additional labels for features within your drawings. No problem, just briefly go back up to Step 2 and find an appropriate place to add them, then come back and move forward. You may also find as you write that an additional drawing might be facilitating. Same thing – if you decide you really need the drawing, just stop, make the drawing (or at least a quick sketch which you can replace with the real thing later), and then come back.

4 Done with that? Now repeat Steps 2 and 3 for each additional set of views you have created that introduce alternate embodiments, extra features, etc. (I didn't promise you you'd only have to do each step once!) Fortunately, after going through the procedure for the first set of drawings, you're probably already getting the hang of it. Anyway, once you've done this for all of your drawings, you're coming close to completion of your specification – and that's certainly something to look forward to.

5 If you've now completed descriptions of the parts of all of your drawings and related functions, now it's time to thoroughly review your document, very specifically asking yourself where can I insert ITBATs ("It is to be appreciated that..." statements) to thwart cheesy opportunistic add-ons. Ask yourself questions like this:

- Is there anything that I can connect to my invention, or variations on what it can connect to?
- Could someone add a generic flow control device at my fluid inlet?
- Can my device be designed to operate at different voltages or powers?
- Does my by device have alternate uses?
- What material substitutions could be made that might affect performance (you do not have to worry about trivial substitutions that have no functional effect)?
- What component substitutions could be made (e.g. a stepper motor in lieu of a servo).

You get the point. Ask yourself these types of questions *until you come up with some*. In most cases, you'll easily generate plenty. Really can't think of anything? Really, don't move on until you do.

6 Final step – don't forget to run a spell check.

Also, did you use any placeholders? I make it a habit when I want to leave a placeholder behind to always include a flag, like the three question marks "???". It doesn't matter what particular characters one uses; what's important is that:

- The flag is something that will never appear elsewhere in the document for any other reason, e.g. don't just use "X".
- The flag is always the same and is attached to every placeholder.

By so doing, once you finish writing, you can always do a final search for the flag to make sure you didn't forget to go back and replace any of them.

Action Step 14

Write the body text of your DETAILED DESCRIPTION OF THE INVENTION under the heading provided in Section 3 of the Application Workbook.

Do not try to insert reference numbers as you write – we will do that as a separate step, next.

8.5 ADDING REFERENCE NUMBERS TO THE TEXT

Introducing reference numbers as cross references will save hours of painful and frustrating manual editing of drawings and text.

Now is the time to start adding the reference numbers. This can be a bit of a chore, and if you try to do this manually, it is easy to miss one and then have to go through and update everything that follows. Fortunately, I am going to show you a method that will save you from that. Instead of just manually typing in numbers everywhere, you are going to insert the number next to the part label as an endnote, just the first time it appears in your text. Each time the part is mentioned thereafter, you can insert a cross reference to the endnote. In this way, if you later realize you missed something and need to add a new identity ahead of parts you have already labeled, just go right ahead and add it with a corresponding endnote, and the other labels will all automatically update.

Since cross-references can be copy/pasted, using them will be just as fast as regular text.

Even better, it's actually quite a bit simpler than that, because endnotes and cross-references can be copy/pasted. So this is going to be just as fast as manually typing the numbers, but you'll finish with a smart document. This is also going to add a handy list of number references at the end of your specification. You are certainly not required to have one, and do not have to include those pages when you actually print the copy that you intend to submit (although I am not aware that there is any harm in it if you want to).

Incidentally, if you really don't want the extra table at the end, you can insert your numbers as SEQ fields in MS Word. However, if you choose this method, you will have to manually bookmark each one to be able to cross-reference to it. Instead, might I recommend just converting the endnotes at the back of the document to hidden text?

Here's the step-by-step procedure:

1 Find your first part label in the detailed specification (you don't need numerical identifiers in any of the rest of the text, although you will occasionally see patents where people did that), add a non-breaking space (Ctrl+Shift+Spacebar), and insert an endnote just after it. Make sure you use the non-breaking space – it's pretty ugly when the part name and the number end up on different lines (but no-one except me will take issue with you over it). When you insert the endnote (pre-Office 2007: Insert\Reference\Footnote; Office 2007 and later: References\Insert Endnote), make sure to select "Endnote" and "End of section". When you do this, Word will jump you down to the end of the section, and you will see a superscript "1". Change this to regular (non-superscript text), add a space or a tab (or a period + space/tab if that's your style) and type the part name, exactly as you always plan it to appear in the document. You can also go up and type a title for the list (this has been done for you in the template).

Now, double click the list number and you'll jump back to the main text where you just entered the end note. The number in your text will be a superscript. Select it and format the font to regular (non-superscript) text. The number should be separated from the text by one space on each side, and should not be bracketed by anything else, no comma's, parentheses, etc. You have just entered your first smart part number.

2 Next use the find function to move to the next instance of your part label. (For those who don't know, since you'll be doing this a lot, here's the quick way: select the verbal part of the label and copy it onto the clip board (Ctrl+c). Open the Edit\Find (pre-Office 2007) or Home\Find (Office 2007+) and paste (Ctrl+v) the label into the "Find what:" window and click go). Once there, add a non-breaking space just after it like before, but now add a cross reference to your endnote instead of a new endnote (pre-Office 2007: Insert\Reference\Cross-reference; Office 2007+: Insert\Cross-reference). Select "Endnote" as the reference type and you'll see a list of your endnotes in the dialogue box (which will be one entry long at this point). In the "Insert reference to:" box, select "Endnote number", click the appropriate end note (see how handy having good names on those endnotes is going to be), and select insert.

3 Ok, so this has been somewhat clunky so far, but we're not going to do that anymore. Now select the cross reference (not the endnote) and copy it onto the clipboard. Open up the find menu again, click "Go" to get to the next instance, enter a non-breaking space, and then just paste in the reference. (It may also have occurred to you that you can select the space along with the endnote when copying to save yourself an operation, but alas, last I checked, if you copy/paste a non-breaking space, the new one comes in as a regular space in MS Word – where do they get these programmers?) Repeat this step until you come to the end of your document. Now that wasn't all that bad.

So why not just use find-and-replace and do it all in one fell swoop? Two reasons. First, often you will find it convenient to have some redundant words in some of your labels. For instance, it is common to have say, a "slider 1" and a "slider latch 2". If you did find and replace, you'd end up with "slider 1 latch 2" every time you meant "slider latch 2", which you don't really want. Admittedly, there are ways around this. You could temporarily find-and-replace every instance of "slider latch" with "X latch", do your find-and-replace to add the numbers to "slider", and then do a find-and-replace to convert every instance of "X latch" to "slider latch 2". But, there's the second reason – MS Word won't cut-and-paste field codes (of which references are a subset). Hey, I didn't program the thing. Take it up with Microsoft.

4 Moving right along, now go back to your first part label number (the endnote) by Ctrl+clicking any of your cross-references. Select and copy the endnote onto the clipboard. Now, move on to your next part label, add a non-breaking space, and paste in the endnote.

If the endnote number comes in with the same value as the last, select it and press F9, or do a "File\Print Preview" (pre-Office 2007) or "Office Button\Print\Print Preview" (Office 2007+) and then exit back to your preferred editing view, either of which will cause the number to update. Double-click the new endnote, and it will take you down to your part label list, where you can format and edit the text to your new part label name. Double-click to get back to where you were, and then repeat the previous steps starting from 2 until you have numerical IDs for all of your part names. For large patents, this can get tedious, but think of doing it totally manually (which I don't think would be much less boring) and then later realizing you need to change something! Anyway, his is pretty brainless stuff, so do can it in front of the TV if it's going to take you more than an hour or so.

5 Ok, glad to be done with that?...Well, you're almost done, but not quite. I recommend now doing searches and applying reference numbers to all pronouns that have part names for antecedents. Hopefully, you took my advice and minimized pronoun use, so this will be quite quick. Common pronouns for which you should search are:

it	its	itself
their	theirs	them
themselves	those	these

If professionals get involved down the line, they may request skip numbering.

Finished? Before we move on, there is one additional minor note worth mentioning. While it is mandatory that part references be numbered in the order they are mentioned, there is no requirement that the numbers be consecutive. As such, many patent attorneys prefer to only use even numbers in an initial filing, anticipating the possibility of having to later add an occasional reference without wanting to change the rest of the numbers. Most users of this guide will be better off drafting their initial filing on their own (but having it professionally reviewed/edited), so there won't be any initial objection to using consecutive numbers, but eventually, if your invention ends up going somewhere, either you or a licensee may get professionals more directly involved, and they may request this. It is, indeed, totally unnecessary because you've generated a smart document where all of the numbers will automatically update if you insert an extra reference, but you will not generally find patent attorneys to be super savvy with respect to the somewhat esoteric inner workings of MS Word, and they may still eventually want this.

Skip numbering is for people who don't know how to use a dynamic numbering procedure like the one above.

You can skip the skip numbering for your application filing.

I don't think it is worth your time to put in skip numbering, anticipating a request they may never come. Nevertheless, I'll show you a way to do this, in case it eventually comes up. MS Word does not allow you to directly specify numbering by two's or anything like that, but below I'll arm you with a method to kluge it if so required without destroying your smart references.

If you must use skip numbering:

1 First, go into Tools\Options (pre-Office 2007) or Office Button\Word Options\Display (Office 2007+) on the view tab and toggle hidden text to be displayed.

2 Now, scroll down to your part label list at the end of the detailed specification section. Double-click the first reference to jump to its location in the text.

3 Copy the endnote onto the clipboard, move the cursor just ahead of the existing reference and paste, which will put a new endnote just ahead of the original.

4 Next, select the new endnote, open up Format\Font (pre-Office 2007) or right click\Font (Office 2007+), and toggle the text over to "Hidden".

5 Double-click to jump down to your part list. You will see a new redundant entry ahead of your original, which you should also convert and toggle as hidden text. Jump back and update references ("Print Preview" or Ctrl+a to "select all" and then press F9.

6 Repeat this for all of your references. Your references are now only even numbers. It's a little klugy (I said it would be), but it should only take you 10-20 minutes, and you can keep the attorneys happy without losing all of your smart references.

Action Step 15 Add reference numbers to the text of your DETAILED DESCRIPTION OF THE INVENTION per the above procedure.

8.6 ADDING REFERENCE NUMBERS TO DRAWINGS

You should complete final drafts of your drawings and insert them into the Application Workbook before proceeding past this point.

You are now going to add smart numbers into your drawings that will be dynamically linked to the text. Before you can proceed with this next step, you must complete final drafts of all of your drawings and place them in the workbook, if you have not already done so. Don't worry, if you later realize you want to make some changes to a drawing, you still can, without having to entirely redo this step, but clearly it will save you rework if you're done editing the drawings before you number them.

To add smart numbers to your drawings:

1 Go to your first drawing and verify that it is pasted in as an in-front-of-text image (not in-line-with-text) or, preferably on a canvas. *Never* mix an in-line-with-text image with floating objects such as text boxes (which is how we will be inserting the reference numbers) – that won't allow you to group objects, which means the relative position of the floating objects to the image won't be stable. Now insert a text box and inside it insert a cross-reference to your first part number, either by the Insert\Reference\Cross-reference (pre-Office 2007) or Insert\Cross-reference (Office

2007+) dialog box or by cutting and pasting from the text. The USPTO requires the number to be at least 1/8th inch (0.32 cm) tall when printed – I recommend 14 font. Now position the text box in your desired location. The number should either be superimposed over the part with which it is identified, in which case (and only in which case) it will be underlined, or somewhere proximal and connected to the part with a line.

Obviously, the number and line should not obscure anything necessary for clear understanding of the drawing, but as you choose a position think ahead to where you will need to place other numbers. Lines connecting numbers to parts within the drawing cannot cross. Clearly, it's easiest within MS Word to work with straight lines, but you can use curved lines if you need to. Numbers count as part of the drawing, so the required page margins apply.

---

2 Inserting additional numbers is quickest by cutting and pasting copies of the nearest text box you have already inserted and formatted using the Ctrl+drag/drop method (just select the text box and drag it to a new location while holding down the Ctrl key – this will make a copy instead of moving the old one. Once the new text box is in place, you can change the cross-reference by inserting a new one over the old one or cutting and pasting from the text.

---

3 Having completed your first image, things will pick up a bit, because to get references into you next drawing, typically most of them can just be Ctrl+dragged from the last image. Indeed, often two drawings (which represent variations on a theme) are so similar that you can take a shortcut by starting by selecting the entire set of number labels from the first image, grouping them, and Ctrl+dragging them down (or cutting and pasting) to your next image as a complete set with only minor subsequent edits required.

When you have completed a drawing, always group all of the numbers with the image (using the "Group" command on the drawing toolbar) to make sure that they stay together, unless you are temporarily ungrouping them to Ctrl+drag/drop part labels. Often, you will find it safer and quicker to make a copy of an image and its labels next to the new drawing that you are numbering, ungroup, move the numbers you want from your copy onto your new drawing, and then delete the copy image and leftover numbers when you are done. This eliminates the risk of your accidentally doing a move instead of a copy on labels from your old figure, often circumvents having to repeatedly switch back-and-fourth between pages if the images are not on the same page, and also avoids simultaneously leaving one figure ungrouped while working on another (which leaves it exposed to all of MS Word's unpredictable quirks). On a side note, if the image part of your drawing is a set of grouped objects, make sure you never simultaneously ungroup your drawing objects at the same time as your text boxes. You'll find it much easier if you keep them as separate groups and ungroup only the one you want to edit.

Action Step 16 Add reference numbers to your drawings as cross references to the text of your DETAILED DESCRIPTION OF THE INVENTION.

94 DETAILED DESCRIPTION OF THE INVENTION

> When you are done numbering your last drawing, this is a good time to scan over your work, just to see what you have accomplished. You are really getting somewhere, and that's pretty satisfying.

8.7 CLOSING STATEMENT

For now, the closing statement to the effect of the preferred embodiments not being intended to limit the scope of the invention is purely a formality.

...but you never know when interpretation of the law will change.

Speaking of getting somewhere, can you believe it's time for your closing statement? Hold your horses, you have a couple of sections left, but this one's a no-brainer. Basically, every patent seems to close with some statement to the effect of the embodiments shown in the detailed description are exemplary and not comprehensive, the extent of the invention being governed by the CLAIMS, blah, blah, blah. You'll find a wide variety of styles, ranging from a single sentence to a moderately long paragraph. The truth is this, since the embodiments in the detailed specification are, by definition, exemplary, and the extent of what is claimed as novel is, by definition, defined by the CLAIMS, the whole thing is entirely superfluous. Still, it seems a little unceremonious to just end the specification with the description of your last embodiment or a string of ITBAT's without some kind of wrap-up, so gin up something…and you never know when some judge is going to reinterpret the law, so an explicit statement isn't a bad idea. Make up your own, or you may use the following verbatim, if you like:

[0062] Closing Statement:

[0063] Having thus described in detail a preferred embodiment of the [INVENTION NAME HERE] of the present invention, it is to be appreciated and will be apparent to those skilled in the art that many changes not exemplified in the detailed description of the invention could be made without altering the inventive concepts and principles embodied therein. It is also to be appreciated that numerous embodiments incorporating only part of the preferred embodiment are possible that do not alter, with respect to those parts, the inventive concepts and principles embodied therein. The presented embodiments are therefore to be considered in all respects exemplary and/or illustrative and not restrictive, the scope of the invention being indicated by the appended claims, and all alternate embodiments and changes to the embodiments shown herein that come within the meaning and range of equivalency of the appended claims are therefore to be embraced therein.

> **Action Step 17** — Enter your closing statement under the heading provided in Section 3 of the Application Workbook.

"The First American" Inventor

It is a shame how little most people know about one of America's greatest architects. Benjamin Franklin was America's first media mogul, developing a printing franchise that extended throughout the colonies. Retiring from active participation in his business interests at age forty-two, Franklin held a number of public offices and played key strategic and diplomatic roles in the French-and-Indian War and the American Revolution, and in the forming of our present form of government. Franklin's greatest fame, however, stemmed from his scientific endeavors related to proving that lightening was electrical in nature, and a number of significant inventions including the lightening rod, Franklin stove, swim fin, glass armonica (a musical instrument), bifocals, and flexible urinary catheter (for his older brother).

Sadly, though indisputably possessed of uncommon genius, Franklin was an "emotional failure," to borrow words Albert Einstein used to describe himself, losing a beloved son to smallpox as the result of what he regarded as a failure on his part to have him inoculated, abandoning his wife (but nevertheless providing for her from a distance in the age-old tradition of wealthy philanderers), and estranging his older son never to speak to him again over his son's loyalties to England during the Revolutionary War – loyalties he had personally instilled. Faults notwithstanding, the influence of Franklin's unmatched body of philosophical, literary, political, philanthropic, scientific, entrepreneurial, and innovative contributions to American culture cannot be overstated.

Claims

9.1 Claim Basics

Exclusively your CLAIMS define the scope of what you claim the right to exclude others from copying without your permission.

Having completed an excellent disclosure of your invention and how it works in your DETAILED DESCRIPTION OF THE INVENTION, only a few formalities are left, right? *Wrong.* Every part of the specification you have completed so far you have done for the benefit of others – that they may know how to reduce your invention to practice. Now it's time to make sure you get your fair reward for so doing, and that is exclusively accomplished in the CLAIMS section of your patent application. With respect to looking after your personal interests from a legal standpoint, this is by far the most important section, and anything that you do not specifically declare in your CLAIMS you have just given away for free. That said, you might expect, therefore, that your CLAIMS be as specific to detail as possible, but quite the opposite is true. As you review claims in your example patents, if they are any good at all, the CLAIMS will seem quite vague. But they are not vague; rather, what they are is quite general...and so should be yours.

Your patent will stop others from making products that contain every element of at least one of your claims.

A nonprovisional application for a utility patent must make at least one claim. Claims protect the invention according to an "all elements rule". Your patent only restricts others from making products that contain every element of at least one of your claims. If your patent claims an invention comprising one or more wheels, one or more motors, one or more pork chops, and a ukulele, anyone on the planet can make and sell something with wheels, motors, and pork chops, and there is nothing you can do about it. Every element you add to a claim restricts or narrows its protection.

Broad claims provide the best coverage, but narrow claims are more difficult to defeat – use both.

Very broad claims are difficult to "invent around" (i.e. to be circumvented by an alternate embodiment of essentially the same invention), but they become moot if anyone can find even a single piece of prior art that would fall under their wide umbrella. Conversely, narrow claims are much less likely to be undermined by prior art, but they are easier to invent around. Therefore, patents generally contain a spectrum of claims, ranging from the most generic that prior art will allow to some that are quite limited but address specific cases of high importance.

Most inventions have several independent claims and many subordinate dependent claims.

Clearly then, the best strategy will be to construct your CLAIMS in layers. At the top level, your claims should be very broad brush, lacking all details altogether, and containing exactly only the minimum essentials to distinguish your invention from prior art, for each novel element individually. These will be your primary or "independent" claims. While strictly speaking, only one invention is permitted per patent application, most inventions actually have more than one independent novel feature (they are essentially several inventive elements that can be used in combination), and so you are allowed to have up to three independent claims for the basic filing fee, and up to two more for an additional $125 apiece. Below these, you will have a larger number of "dependent" claims, which reference parent claims (which can be independent claims or other dependent claims) and then introduce additional details, one at a time, again in the most general and broadly encompassing terms possible.

The additional details of the dependent claims effectively narrow the scope compared to the corresponding unconditional primary claim(s), in order to:

- Provide a slightly narrower, backup claim, in case the independent claim is ultimately defeated in litigation.
- Introduce additional novel features that you specifically want to protect when used in conjunction with the independent claim. Note that if these novel features can be employed in abstraction from the part of your invention covered in the corresponding primary claim, then they should also be called out as independent claims.
- Combine two independent claims so that the composite is specifically covered.

Dependent claims can add to, but not take away from, parent claims.

Dependent claims cannot contradict or arbitrarily modify a parent claim, but can only add additional elements or further constraints to its elements. Clearly, then, dependent claims are always more narrow than their parent claims. That's a pretty straightforward concept, but a few bad examples will probably be helpful:

1. A wheelbarrow comprising:

    a) a bin with;

    b) a spout;

    c) at least one wheel; and

    d) at least two handles.

    Here's an independent claim for a wheelbarrow with a pourspout on the bin to facilitate more precise off-loading of the contents.

2. The wheelbarrow of claim 1, but without the wheel.

    This claim is invalid. Dependent claims cannot subtract elements. Claim 1 should have left off the wheel, which should then have been added in claim 2.

3. The wheelbarrow of claim 1 where the wheels are replaced by skis, such that the wheelbarrow may easily be moved over snow.

    This claim is invalid. Dependent claims cannot make element substitutions. The correct strategy would be to start with a more general term for wheels, skids, skis, etc, and then call out wheels and skis (and other options) in dependent claims.

4. The wheelbarrow of claim 1 with at least one handle.

    This claim is invalid. I already said at least two handles in claim 1, so I have these backwards. Claim 3 contradicts claim 1. I need to write "at least one handle in claim 1", and then "at least two handles" in the dependent claim if that is my intention.

5. The wheelbarrow of claim 1, where the bin is made of polyethylene.

6. The wheelbarrow of claim 5, where the bin is made of a polymer.

    Claim 5 is fine, but claim 6 is invalid. Claim 6 references claim 5, but calls out a less specific material for the bin (polyethylene as a particular type of polymer), so claim 6 is more general than claim 5. If claim 6 had referenced claim 1 instead, that would have been OK, although better form would be to switch the order of claims 5 and 6, so that the "polyethylene" claim references the "polymer" claim as its parent.

So the claims in your patent essentially represent a multi-layer defense against would-be attackers. If one of your most general (independent) claims falls in battle, then one of your secondary, slightly more specific claims kicks in, and you haven't lost the whole war. If that claim falls, another one backs it up, and so on...up to a limit. For very logical reasons you are allowed many more dependent than independent claims, up to a maximum of twenty-five total claims (meaning twenty to twenty-four dependent claims, depending on the number of primary claims). Incidentally, the upper limits on purchasing additional claims are actually a rather recent (37 CFR 1.265, effective November 1, 2007) change from the old rules, where you could pretty much just purchase as many extra claims as you wanted for very minimal fees. This is important to note so that you are not fooled by the many patents filed before 37 CFR 1.26 that have more than twenty-five claims.

Think of your CLAIMS as a multi-layer defense.

In the spirit of the new legislation, the USPTO has also become more diligent about enforcing the one-invention-per application rule, where, if your filing contains claims pertaining to two distinctly different inventions (instead of different aspects of the same invention), you will receive what is known as a "Restriction Requirement", which means the patent examiner will divide your claims according to which invention they pertain and ask you to select the group you wish to keep. The set you do not select will be dropped from the application, but are not lost – as they have already been ruled a second invention, you may resubmit them in a "Divisional Patent Application" if you are up for it.

You can claim only one invention per patent application.

You can't get around the claim limit rules by simply filing multiple patents, by the way, unless they really are distinct inventions. Specifically, the claim limits apply jointly to all co-pending applications by the same person or organization that have at least one "patently indistinct" claim. Frankly, I think the new system is better from a practical standpoint – you get your three to five best independent and the balance on twenty-five best dependent claims to carve out what's yours, and if you can't fit it in a scoop that size, you're probably trying to take too much (but there is an argument that inherently complex fields of invention are disproportionately impacted). Functionally, limiting the number of claims keeps patent processing fees within reason.

You cannot circumvent claim limit rules by filing multiple patent applications unless you really can partition your invention into two that are completely distinct.

Of course, there is one loophole (this is, after all, government we're talking about). If you want to exceed the claim limits, you can file an "Examination Support Document" (ESD), which reports:

- A detailed search of US and foreign patents and non-patent references
- The search resources and strategy
- A thorough analysis of the claims of each reference produced by that search comparing them to the claims of the patent application
- A statement of what the invention is and why it is patentable (back to novelty, non-obviousness, and usefulness).
- If you do not qualify as a "small entity", you would also have to map the limitations of all your claims specifically back to the extant claims in the prior art. If you're a small independent inventor, you definitely qualify, but you'd have to file a waiver to that effect.

...of course, there is a loophole, if you have the cash.

For now, consensus seems to be to stay away from ESD's.

Getting all that done is very expensive, and, even if you did, it creates a weakness by adding a second avenue of attack for your claims. If you opponents cannot find a crack in your claims, they may be able to instead prove your ESD incomplete or invalid, sort of like knocking the foundation out from under an otherwise impenetrable fortress. Incidentally, the new limits were challenged in court and upheld in a March 2009 ruling. There is some conjecture that the new rules may still eventually be overturned, but at the time of this writing, the general consensus in the legal community seems to be don't go over the five independent/twenty-five total claim limit.

If you receive no objections to any of the claims filed with your original application on the basis of their being too general, you've probably been too narrow.

Moving on, you should be deriving from the above discussion that crafting claims is all about strategy – almost like a game, except the stakes are real. You must be specific enough to distinguish your invention from prior art, but only exactly that restrictive and no more. It's often said that if your first office action comes back without objections to your initial draft of claims on the basis of their being too broad, you haven't claimed enough. Also, keep in mind that establishing the CLAIMS that actually end up in the final patent is very much a negotiation, and, as in all negotiations, you should make sure you put something in you plan to give up – the other side always needs to be able to come away having done their job.

9.2 THE STRUCTURE OF CLAIMS

Each claim should form a single (sometimes rather long) sentence.

Your CLAIMS section must be on separate pages from everything else. Start the first page with the heading "CLAIMS", followed by the words "I claim:" on the next line. Below this your claims will read as a list of ways to complete the sentence "I claim..." Do not write the "I claim" over and over again; the colon after "claim" means it already applies to every line-item in the list. Claims must be numbered in consecutive order, and we will do away with the [XXXX] paragraph numbering to avoid confusion. The text of each claim must start with a capital letter and end with a period, and may not contain any other capital letters or periods. Clearly this excludes multi-sentence claims, but you may have as many clauses as you like, so don't worry about run-on sentences; these will be king-daddy run-ons. It will be best (for you and the examiner) to organize each claim using an outline format, but do not be fooled into thinking the claim is an outline. It is, indeed, the completion of a sentence starting with the words "I claim", so every clause of the claim must be punctuated to distinguish it from the next. Use semicolons, except for on the second to last line, use "; and" and, on the last line, use a period. Here's an example:

CLAIMS

I claim:

1. A food product comprising:

   a) a pastry;

   b) a stick providing a means of holding the pastry; and

   c) a removable wrapper surrounding the pastry.

Outline formatting is optional, but recommended.

Notice the outline indentation and letter numbering of the subclauses. These features are allowed but entirely optional. There are no specific rules regarding the indentation amount or numbering/lettering scheme for the outline subsections. The USPTO states a preference to have the indention, though, so I recommend you do that at least. The number at the very beginning of the claim is mandatory, however. Regardless of whether or not you use outline formatting, it is best to place each subclause on its own line.

Typically, fewer words means more general.

Ordinarily, the simpler each subclause, the better, because that usually implies more generality. More words make the clause more specific. You should only use as many words as necessary to distinguish your invention from any prior art, except, of course, as you move down the list your claims will incrementally add details. We might continue this notional list by adding a number of dependent claims, as follows:

---

1. A food product comprising:

   a) a pastry;

   b) a stick providing a means of holding the pastry; and

   c) a removable wrapper surrounding the pastry.

2. The food product of claim 1 where the pastry is circular, having a hole in the center.

3. The food product of claim 1 where the pastry is a doughnut.

4. The food product of claim 1 where the pastry is at least partially frosted.

**Parent claims
+ Additional details
= Dependent claims**

Note that claims 2 through 4 reference claim 1, but add additional details. In the event that someday someone challenges claim 1 by citing some 12th century etching showing feudal peasants gnawing bread off of sticks because their hands are filthy and they can't afford utensils, claim 2 will stop the run, as long as the picture doesn't show holes in the article being consumed. Claim 3 essentially says the same thing but pulls in the strength of all that exists in the art regarding what defines a doughnut, but also incurs the associated definitional boundaries. Claim 4 adds in the option of frosting, which adds another layer of defense to the basic invention (since few people will want an unfrosted doughnut-on-a-stick*), but also specifically covers a derivative option.

But what about powdered sugar coatings? Better revise that one, or add another claim to cover it – these are the kinds of questions you should be asking yourself constantly as you write your claims.

We can keep going. We can write dependent claims that use other dependent claims as their antecedents, such as:

5. The food product of claim 4 with decorative sprinkles.

You should perpetually think about how you would attack your CLAIMS.

If you don't find the weaknesses, someone else will.

Constantly ask yourself devil's advocate questions as you craft your CLAIMS. For instance – as I wrote claim 5 above, I was tempted to write "edible decorative sprinkles" instead of just "decorative sprinkles"...but, then I thought about it and realized I want to claim sprinkles edible or otherwise. I can't think of why anyone would even think of using inedible sprinkles (think of the liability), but why should I place arbitrary limits in my claim? I shouldn't. If these claims were attached to a real patent I was writing, I'd have a very hard look at the adjective "decorative". In fact, I'd delete it, but only after I went back to my DETAILED DESCRIPTION OF THE INVENTION and made sure "sprinkles" were well defined in type and general purpose, but without imposing any limitations. *Beware of adjectives in your claims.* Never add an adjective without a very specific and conscious intent (more on this later).

Let's go back and look at the elements of that original independent claim:

a) a pastry;

Pastry's pretty general, but could I be more general? It comes down to trying to claim everything that does not exist in the prior art, up to the point of not being omnibus – the patent examiner's job is to prevent me from claiming so broadly that I end up owning later inventions to which I have no legitimate entitlement.

b) a stick providing a means of holding the pastry; and

Does it have to be a stick? Could I have just said "handle"? Does the stick have to provide a means of holding the pastry, or could I just leave that clause out? Again, it comes down to trying to claim as broadly as I can, but still be specific enough for the claim to be considered to encompass a single invention. Because you're allowed multiple independent claims, you don't have to claim it just one way. I can call it a 'stick" in one claim, and a handle in another. I can call it a handle in an independent claim, and then specify the sub-case where the handle is a stick in a dependent claim. Similarly with spelling out the purpose of the stick – I can leave that out and bring it back in a dependent claim.

Be careful with purposes. Sometimes (especially on such a simple invention) they may be necessary to clarify the scope of the invention to something the patent examiner will allow, but anyone can poke a stick into a doughnut and potentially say it serves a different purpose. Of course, a court may derail such a ruse, but, then again, it may not.

c) a removable wrapper surrounding the pastry.

Should we call it a "removable wrapper" or just a wrapper? Clearly, I'm less likely to have my claims rejected if I'm more specific about the intent. But is "removable wrapper" really more specific than "wrapper". I'd argue that, in this case, I've only presented the illusion of being more specific (a good thing), because "removable" is what I call a guaranteed absolute. Even if I made the wrapper out of three-foot-thick reinforced concrete I could still technically remove it (with the help of jack hammer), and, after all, Princess Leia got Han Solo out of a solid block of "carbonite" with the touch of a button. More to the point, the word "removable" illustrates intent without narrowing scope, and so it's definitely a keeper.

Here's a claim that we might want to write:

5. The food product of claim 3 or 4, with sprinkles.

Traditionally most applicants have avoided multiple dependent claims because they cost extra.

This is a perfectly legitimate claim, and is known for obvious reasons as a "multiple dependent claim". You won't see a lot of them in the patent literature and I can tell you one very good reason – they cost extra. Not a little extra, but previously on the order of 50% of the basic filing fees, EACH. More typically, in the patents you review you will typically see two regular dependent claims used in the place of one like the above example; one referring to claim 2, and one to claim 3. Clearly as layers of claims build up, this can add up to quite a little network of parents and dependents; e.g., if I have three dependent claims referencing an independent claim, and then want to claim another two dependent features that work with those three dependent claims, for those features I'm adding $2 \times 3 = 6$ new claims. Prior to the new law limiting total claims to twenty-five this was a non-issue, since extra dependent claims beyond the number that came with the base application fee could be had in unlimited number for $9 apiece. The revised rules don't make multiple dependent claims any more useful than before, however, as each dependency is counted as equivalent to one regular dependent claim – so they do nothing to help with the twenty-five claim limit. Also, claims which are dependent upon a multiple dependent claim are also counted as an equivalent number of claims equal to its dependencies; thus, multiple dependent claims really don't provide a loophole (the only one being the formidable ESD).

With respect to claim limits, each dependency of a multiple dependent claim counts as an additional claim.

Multiple dependent claims can only be used in the alternative, by the way, not the conjunctive. That means claim 6 above is OK, but I can't write:

5. The food product of claims 3 and 4, with sprinkles. ☒ **No**

The difference between claims 6 and claim 7 is purely a word choice. They essentially mean the same thing, but the former would be allowed, while the latter would be equivalent to writing "Dear patent examiner, I don't know what I'm doing".

There are circumstances where the matter doesn't amount to a simple word choice. Say I added to my claims list a second independent claim describing a novel stick that had barbs keeping the doughnut from falling off. I might claim the stick in claim 8, and want to write for claim 9:

9. The doughnut of claim 1 using the stick of claim 8. ☒ **No**

Multiple dependent claims cannot combine parent claims.

Multiple dependent claims cannot reference other multiple dependent claims.

You can't do this. Multiple dependent claims must reference the prior claims using "or", "any one of", "one of", "any of", and/or "either". They cannot combine two previous claims. The correct way to structure claim 9 above would be to start out with "The doughnut of claim 1 using" and then add the full text describing the stick of claim 8, with no actual reference to claim 8. That's always seemed silly to me, but that's the way it is.

Also, so that patent examiners don't develop nervous ticks or do bad things to themselves, multiple dependent claims are not permitted to refer to other multiple dependent claims, either directly or indirectly.

9.3 Qualifying Language

Qualifying language tailors claims to be not overly general and exclude prior art.

Use it carefully.

Now, our doughnut-on-a-stick is a very simplistic invention. There is minimal qualifying language detailing the function of the elements or their relative relationships, and the independent claim stakes out almost anything and everything comprising the three basic elements (a) pastry, b) stick, c) removable wrapper. The example does include some "qualifying language", however, in the form of the phrase "providing a means of holding the pastry" and the word "removable". The first of these very specifically narrows the scope of the invention, whereas, as I have previously pointed out, the second does not, but both qualify the list of elements by bringing intent into the picture. Whereas most of the examples to follow contain little or no qualifying language for simplicity (they illustrate other aspects of claimsmanship), qualifications are almost always necessary. But didn't I just say we want to be as broad as possible? Yes, but only to a point. We must also be sufficiently narrow (and no narrower) to limit the scope of the CLAIMS to be valid in the context of prior art. If any of your claims could be equally applied to another existing invention, it's the patent examiner's job to find out about it and deny that claim.

There are four ways to limit your claims. You can:

- Add additional required elements.
- Place extra constraints on the elements.
- Place constraints on the way the elements interact (usually preferred) or interconnect (usually not preferred).
- Add language detailing intent.

You should come up with at least three independent claims for your invention.

Generally, it will be best to use combinations of these, always with the goal to maximize generality while just barely disqualifying yourself from prior art. Remember that you get multiple independent claims – if you are torn between two strategies, your patent will be stronger if you implement both. Since the price is the same for one to three independent claims, you should come up with at least three (but three is generally enough; I wouldn't pay for more unless I had other content I needed to claim that I couldn't accomplish within the first three).

Qualifying language can be integrated into your elements, like so:

> 1. A leg comprising:
>
>   a) a thigh bone;
>
>   b) a knee bone connected to the thigh bone;
>
>   c) a shin bone connected to the knee bone;
>
>   d) an ankle bone connected to the shin bone; and
>
>   e) a foot bone connected to the ankle bone.

Integrated qualifying language works best when the inter-relationships between elements are simple.

OK, I did say that it's usually better to speak to how the parts interact instead of how they interconnect, so if you noticed, that's good – you're paying attention. The real point of this example is where the qualifying language (the "connected to" clauses) goes. Notice that each added element only refers to already-stated elements. I'm not aware of any rule forbidding referencing forward (after all, outline formatting is optional), but it's bad form and defeats the intent of using an outline format.

Alternatively, qualifying language can come after the elements list:

> 1. A leg comprising:
>
>   a) a thigh bone;
>
>   b) a knee bone;
>
>   c) a shin bone;
>
>   d) an ankle bone; and
>
>   e) a foot bone;
>
>   where the knee bone is connected to the thigh bone, the shin bone is connected to the knee bone, the ankle bone is connected to the shin bone, and the foot bone is connected to the ankle bone.

Complex inter-relationships are better off segregated from the listed elements.

This claim is entirely equivalent. In this particular case, I think it's obvious that the first example works better, since each element is qualified only once and directly to one other element. The second form, however, will be more convenient where the qualifying language refers more globally to the elements, such as describing the action of a complex assembly. You may, of course, use a combined format, where direct inter-part relationships are embedded in the element introductions, and then more global qualifications placed at the end.

9.4 TERMINOLOGY

Seemingly innocuous wording choices can make a vital claim worthless.

Now that you understand basic claim strategy, I need to warn you there are some very specific linguistic booby-traps in the law which you simply need to know about and avoid. The turn of a tiny phrase can make all the difference in the world for the interpretation of a patent claim, and, guess what – the legal definitions of words continuously change. Scary, huh? I have no remedy for you on this one. When a change is made, the information is posted on the USPTO's website (usually in less that straightforward legalese, unfortunately), so if you keep up with the news, you could stay abreast of it, but, that's a lot of effort, and unless you're a full-time professional in the field, that probably won't be very practical.

Having an attorney review your claims is strongly recommended.

If you've applied the lessons of this guide well, attorney fees will be minimal, simply because there won't be much to do.

So, while with this guide I'm going to endeavor to arm you as best I can to make a good go of it solo, I'll tip my hand that at the end of this section I'm going to turn on you and encourage you to involve a patent attorney; but, believe me, if you've effectively utilized this guide his/her role should be minimized to simple editing and refereeing your word choices, and that will cost you almost nothing compared to what otherwise would be the case. If you truly have applied what you've here learned (and you should be able to approximately gauge your competence level), in the event that an attorney reviews your work and starts into some spiel about how your initiative is not going to save him that much time (and you that much money), find another attorney.

9.4.1 Using Consistent Terminology

Every element in your CLAIMS should always be referred to using exactly the same words.

Your CLAIMS should not include reference numbers.

Just as we discussed in Section 7.2.2 in regard to the body text of your specification, standard creative writing style rules won't apply for your CLAIMS. Do not find creative alternative ways to refer to a single element in your invention. You must call each element by exactly the same name every time, and those names should be the same as in your DETAILED DESCRIPTION OF THE INVENTION. You do not need to, and certainly should not, reference every numbered element, as the preferred embodiment will contain numerous details that are not definitional for your invention. In your CLAIMS, even elements that you numbered in your drawings that you do call out should not actually include the reference numbers, as this is overly restrictive. Some of these elements may not be numbered elements at all, but you can still make sure to call them by the same name in your claims. Since your DETAILED DESCRIPTION OF THE INVENTION serves as the key to understanding the rather vague terms of your CLAIMS by example, it will be in your best interest to make the correspondence between the two clear and not easily subject to twisted interpretations.

9.4.2 Connecting Words

Above all else, make sure you understand the legal distinction between "comprising" and "comprised of".

From the above discussion you should be gathering that word choices are very important as you write claims. It's not really all that bad, but there is one word choice rule to which you must, *must* adhere, and this one won't raise any objections from your patent examiner whether or not you make the right choice. Do you know what the difference is between "comprising" and "comprised of" is? (If your patent turns out to be worth $100M, then there's a good bet the answer is $100M.) These are what are known as "connecting words" or "transition words" – they are the words that connect a claim element and the list of things that make up that element. To the USPTO, there is actually quite a difference between these two expressions.

"Comprising or comprises"

An Invention A comprising sub-elements b, c, and d is defined to have these sub-elements as a minimum, but other things may be included without altering the inventive concepts embodied therein. Likewise if it is said that Invention A comprises sub-elements b, c, and d.

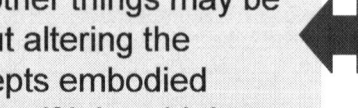

"Comprised of"

An Invention A comprised of sub-elements b, c, and d is defined to have ONLY these sub-elements. A device including an additional feature would not be covered, i.e. all someone would have to do is add one new feature and you couldn't touch them.

Always use "comprising" or "comprises"!

Clearly you should always use "comprising" or "comprises", and never use "comprised of". Are there other preferred connecting words? No. Here's a list of the problems with other common connecting words:

- Including
- Containing
- Characterized by

These are technically equivalent to "comprising" but not recommended, because comprising is simply standard at this point. You never know when some judge is going to hijack these less precise terms. When you mean comprising, just say comprising.

- Consisting of

Essentially equivalent to "comprised of", so absolutely do not use.

• Consisting essentially of	The addition of "essentially" attempts to convert "consisting of" to something like comprising, but doesn't quite get there. An Invention A "consisting essentially of" sub-elements a, b, and c means that the only important elements of Invention A are sub-elements a, b, and c; minor additions are included; but an invention with any substantial addition would not infringe, making "comprising" much superior.
• Having • Composed of	These connecting words are ambiguous and are "interpreted in light of the specification" per the USPTO. Clearly it's better not to use claims that are open to interpretation – look how well that's worked for the U.S. constitution.

Yes, there are booby-traps in patent case law.

Since intellectual property disputes can involve substantial sums of money, you should expect it to be dirty business.

At the end of the day, I don't know how many different ways to convince you to use "comprising/comprises" instead of anything else, but no number would be too many if that's what it took. Why is it like this? All I can suggest to you is to consider *cui bono*? The distinction in interpretation of connecting words is literally a booby trap – and without a doubt there are some bad people somewhere in the past who created and exploited this unsavory loophole to cheat others. Clearly, the fairest thing would be to interpret all connecting words as equivalent, since that's obviously the inventor's intent (it's pretty plain that no inventor filing a patent would deliberately arbitrarily limit the scope of his claims, so anything of the sort is overtly accidental). Moreover, the existing art readily serves the purpose of defining necessary limitations for the interpretation of connecting words in later patents (i.e. connecting words may readily be assumed to be intended to be as broad as possible, save where they would infringe prior art). Maybe in the modern USPTO's excellent spirit of encouraging small inventorship this will eventually change. For now, don't be a victim.

9.4.3 Adjectives, Etc.

Your adjectives should never narrow the scope of a claim.

Frankly, with regard to the use of adjectives in your independent claims your goal should be effectively none. Now, strictly speaking, you'll be hard-pressed to use absolutely no words that linguistically qualify as adjectives, but adjectives effectively can serve two purposes, and you should learn to distinguish them. You can, and sometimes pretty much must, use adjectives as part of an element name, but you should never use them to constrain anything about the character of the element to which they are attached. In our example claims above, you will see the use of the label "food product", where "food" is clearly an adjective. This is OK, because I'm genuinely describing something I'm referring to as a "food product" where the two words form a name. The two words act as a single compound word, but in English it just happens that history has never specifically combined them. I cannot make my meaning clear by simply stating "product".

Real qualifying adjectives belong in the DETAILED DESCRIPTION OF THE INVENTION, but still keep the language as exemplary, not exclusive.

This is not the same as in our example claim 5 on page 102 where we called out "decorative sprinkles". Here the element "sprinkles" is adequately defined on its own without the adjective "decorative". Adding "decorative" unnecessarily narrows the scope of the element. It is only in the DETAILED DESCRIPTION OF THE INVENTION where it will be helpful and wise to provide additional details in regard to the intent of the sprinkles, because that section serves to illustrate and support, but technically not narrow your claims (even still, since the remainder of the specification is used to interpret the CLAIMS, best practice is to use open-ended language rather than closed – your specification can state decoration as an example reason for the sprinkles, rather than *the* reason). Use adjectives as part of a name for an element, but never as a descriptor that narrows the scope.

Whereas narrowing a claim element is one of the options listed on page 97 to avoid overlap with prior art, this should generally not be attempted by adding a qualifying adjective for two reasons:

1. If the prior invention's claims are well drawn, simply adding an adjective cannot move your claim outside of the scope of the existing claim, since doing so only reduces scope, but does not affect intent.

2. If the prior invention's claims are flawed, containing, say, an extra adjective or two that unnecessarily leaves an opening that really does allow you to dodge them with the simple use of an adjective, I'll re-iterate my strong loathing of those who take advantage of others' mistakes. Moreover, doing so will weaken your patent by inviting others to do the same, and ultimately, the original patenter could file a continuation to correct the error, in which case you might be out of luck (depending now only on your legal horsepower). If your invention is novel, useful, and non-obvious, you will be able to craft a quality set of legitimate, non-predatory claims.

Think carefully when choosing compound names for elements.

Here's a common example: Consider the word "cylindrical". Lots of components of many devices tend to be cylindrical as their most logical and natural shape. So when one calls out, say, a piston in a claim, there is a very natural tendency to say "cylindrical piston"…and it sounds so much cooler. Admittedly, alternate geometries for pistons haven't made a big splash and aren't expected to do so anytime soon – so calling the piston out as cylindrical is probably not a big deal right? Well maybe, but why arbitrarily limit your claim? In your claim, simply call it a piston. In your DETAILED DESCRIPTION OF THE INVENTION, you can go ahead and point out that the preferred embodiment is cylindrical, but that just means it's preferred right now. If, two years later, someone discovers counterintuitively that ovular pistons wear half as quickly, that'll be OK, because your claim covers your invention employing any piston geometry. This isn't quite as far-fetched as it may seem at first. Remember when elliptic bicycle sprockets became the next thing? Hope all those guys holding prior sprocket-related patents weren't cavalier with their adjectives.

Adjectives which represent important beneficial qualifiers should be introduced in dependent claims.

Now, on the other hand, if the adjective in question genuinely improves the invention, clearly you do want it in your claims; but, instead of limiting your independent claims, move it to a dependent claim. One good example would be to claim a natural and difficult-to-avoid material choice for a particular element of the invention. Do this only where the adjectival property would not qualify as obvious. In our piston example above, it will be obvious to anyone skilled in the art that a piston could be cylindrical, so adding a dependent claim of the form "The thingamajig of claim X where the piston is cylindrical" would simply be a waste.

Avoid Material Callouts

Avoid talking about specific materials unless the material choice really constitutes a fundamental part of the invention.

Another thing to watch out for is the use of words describing material choices. Your claims pretty much should never include any material specification, unless it is the material you are actually claiming. In most cases, these are simply details that come into the picture when you describe a preferred embodiment in your DETAILED DESCRIPTION OF THE INVENTION. Don't even describe properties of the materials, if you can avoid it. For instance, if I need a stopper in a flask, I could call it a latex stopper; but realize that more generically I should call it a rubber stopper; but, then again, more generically still can I not simply call it an elastomeric stopper? In the end, why not just say stopper?

Avoid Specific Quantity Callouts

Seldom should you specify a particular number of anything.

There are only two numbers of relevance to a patent claim, one, and more than one. For instance, if I have bleed holes in the above-mentioned stopper, does it matter how many? Generally no, so 99.9% of the time, when you refer the quantity of anything you should be saying "one or more", one or a multiplicity of, "one or a plurality of", "at least one", etc., all of which mean my claim covers the invention with any number of the particular element in question. Once in a great while, it will be necessary to have exactly one of something, or some minimum larger than that, etc., and then you can be more specific, but be very cautious if you believe you're in that category. I expect if you think about it a little longer, you'll find that you're not.

Do not include technical details necessary to make your invention work. Include only details necessary to uniquely define your invention

Do not confuse technical necessities with conceptual ones. Your invention may only work if a particular solution is between 1% and 3% nitric acid, but that doesn't mean anything for your CLAIMS. You should just call out a nitric acid solution, unless the limits are necessary to avoid infringing on an existing patent (which probably implies the guy who wrote the original patent blew it by not being general enough). That technical detail may, however, make for an excellent reinforcing dependent claim.

Avoid Specific Dimensional Callouts

Dimensions have no place in patent claims.

Dimensions, have no place in patent claims under any circumstances. If there is something particular about the length, height, volume, etc. of some element, specify it in terms of intent. There are no absolute quantities of importance in the universe, and that isn't going to change for the purposes of writing your patent. Relative measurements are all that matter. If you have two pushrods where it is important that one be longer than the other, just say one is longer than the other. Don't put down how long they are or how much longer one is than the other. Even if you think it is necessary for one to be longer than the other, you are usually better off not stating that in your claims. Generally, there won't be any reason to.

Avoid relative size references in favor of simply describing inter-part relationships from a functional perspective.

Another example: if you have a pair of concentric axles where one rotates inside the other, you can just state exactly that; you do not need to specify that the inside one is smaller in diameter that the outer one. That's implied, but stating it just opens the door for opportunists. What if someone just makes the internal shaft bigger than the outer shaft elsewhere along the length, where it is no longer constrained? Depending on your wording, the extra (pointless) adjective could spell doom. If you have any adjectives that address size, or even relative size of elements of your invention, work hard to remove them.

Avoid Specific Positional Callouts

Describe the interaction of the parts of your invention instead of their relative position.

In the same spirit as adjectives, prepositional callouts to relative position also weaken patent claims. For instance, in a patent claim, unless gravity is involved to move something in a particular dimension never use "above" or "below" or any reference to orientation. Never use "in front of" or "behind". In fact, don't use even the generic "adjacent". When it comes down to it, the position of the elements of your invention really isn't important. If a bumper limits the travel of something, just say that. If a hole allows flow from one cup to another, just say that. State only how the invention elements relate to, or interact with, one another. You really don't need any specific language about position, and such language is very, very dangerous. The easiest way someone can make your invention without permission is to make trivial changes to the layout if you were foolish enough to make your claims configuration specific.

All of which brings us back to...

> **With regard to the use of adjectives in your independent claims your goal should be effectively none.**

Do unto others as you would have them do unto you.

Frankly as a real inventor, I have to regard you as very low form of life if you are, indeed, trying to cheat someone out of their invention on some technicality they overlooked in their claims – no matter how poorly their patent claims are written. It's no crime to not be an expert at writing patents, and unfortunately few inventors are. My goal with this writing is to help you protect what's yours; but clearly, a full understanding of patent claims can equally be used to take advantage of others who are not experts. Even if it is a virtual certainty that someone will exploit that flaw you find in what turns out to be a valuable patent, it doesn't have to be you. Please work hard to

Scrub your claims thoroughly to iron out extraneous qualifiers.

keep what's yours, and leave to others what's theirs. What you can do for yourself, once you have drafted your claims, is to diligently comb through them and iron out extraneous adjectives and other unnecessary qualifiers. You'll find some, and you'll be glad to be rid of them.

9.4.4 Avoid Absolutes

Do not claim conceptual ideals that cannot actually be attained.

While avoiding unneeded adjectives will keep you out of much trouble along these lines, the rest of the English language contains much that can still foil your claims. It is also imperative that you write in attainable instead of ideal terms. What does that mean? It means, for example, that the new kind of fan you've invented isn't "silent", but rather has "reduced noise" or its elements "limit" the noise, or "abate" the noise, or whatever, so long as you do not claim it is absolutely noiseless. Why? Because it can't really be completely silent, and if you claim it is, you're claiming something you haven't actually invented. "Silence" is a word we use to describe a concept, but it doesn't actually exist as a state in the natural world. Surely when you envisioned it, you were determined to figure out how to make a fan that was "silent", but you really made one that has "reduced noise". Whatever problem your invention is intended to eliminate, be careful – always ask yourself does it truly eliminate it, or merely limit or reduce it?

9.4.5 Avoid Alternatives

Use a generic term instead of calling out alternatives in a single claim.

Some patent examiners will object to the use of alternatives within a single claim, and I expect with the new claim limits this will be even more the case if you are maxing out your claim number and trying to use the word "or" to get a two-for-one. This is a simple one to avoid. Instead of calling out alternatives, use a more general term that encompasses both. If the particular alternatives are important, put them in dependent claims.

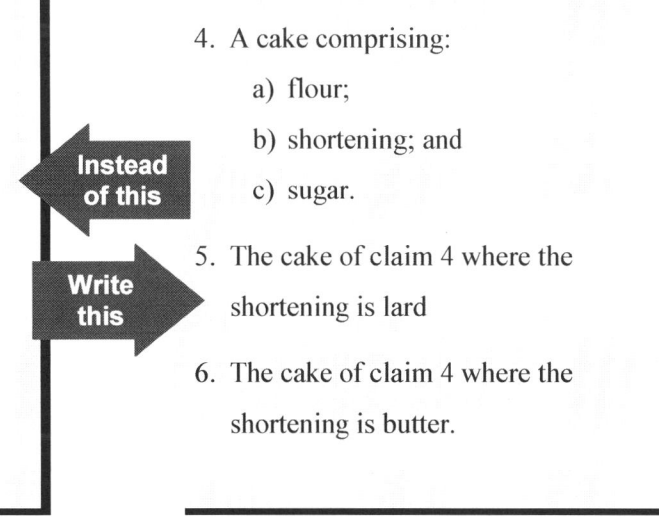

4. A cake comprising:
   a) flour;
   b) lard or butter shortening; and
   c) sugar.

Instead of this

Write this

4. A cake comprising:
   a) flour;
   b) shortening; and
   c) sugar.

5. The cake of claim 4 where the shortening is lard

6. The cake of claim 4 where the shortening is butter.

9.4.6 Means-Plus-Function Language

Means-plus-function language is an excellent example of a once useful tool that lawyers and judges wrecked for motives not readily discernible.

Here's a good example of terminology that changed in definition recently, capriciously, and for the worse. You will see very commonly in older patents, and patents that aren't all that old at all, something called "means-plus-function language". Means-plus-function language states an element in terms of its intent rather than specifics, e.g. for an electrical device, instead of creating a zillion claims calling out different power sources such as plugged in a wall socket, using a battery, generator-powered by a water wheel, etc., you could just call out a "means for supplying electrical power", and that would pretty much cover everything relevant. Traditional patent law provides very intentionally and specifically for the use of such language simply because it presents a means of being very concise, general, and unambiguous at the same time by allowing one to describe an element in terms of what it does, instead of by its physical specifics:

> TITLE 35 > PART II > CHAPTER 11 > § 112)
>
> "An element in a claim for a combination may be expressed as a means or step for performing a specified function without the recital of structure, material, or acts in support thereof, and such claim shall be construed to cover the corresponding structure, material, or acts described in the specification and equivalents thereof."

Technically, it's still safe to use means-plus-function language if you call out every intended means in the specification.

Think of how useful this is (or was). Consider just this particular example – think of how many different patents are out there teaching electrical devices, but for which the particular source of electrical power is arbitrary. Until recently, you could just call out "a means for supplying electrical power" and move on to the features of the invention that are actually important. Then, suddenly a court decided that the "means" in means-plus-function only applies to means specifically called out in the patent specification. I think you'd have to live off tax dollars to understand why, but, whatever the "reason" (maybe "motive" would be a better word?), Presto-chango and "Poof!", what was a very useful and sensible structure to prevent cheesy attacks on patents by making unimportant changes to tertiary elements was gone.

If you really can't avoid means-plus-function language in your CLAIMS, the next best thing will be to specifically state that the means is arbitrary in your specification.

Now, generally it will be best to avoid the use of means-plus-function language unless you really plan to and can list every possible permutation of "means" in your specification. With respect to our electrical power source example, you'd be best served now to avoid mentioning the power supply in your independent claims at all, simply limiting the claim to the rest of the device, and then perhaps bringing in the means of supplying power clause in a dependent claim. To be thorough, if you do end up with means-plus-function language in your claims (and I really expect there will be cases where it'll be next to impossible to avoid), may I recommend making sure to specifically include in your DETAILED DESCRIPTION OF THE INVENTION (in addition to listing the most obvious and important examples) an ITBAT to the effect of:

> [0058] It is to be appreciated that [specify means] is arbitrary and many possible substitutions will be apparent to those skilled in the art which do not alter the inventive concepts embodied in the [Invention Name] of the present invention".

Claim and specification language must continuously evolve with the law.

This language is untested (so use at your own risk), but that's how the patent world evolves. Patenters and attorneys come up with protective language. The rules change (corruption and incompetence being the primary culprits), and patenters and attorneys have to do their best to invent new language that forms an effective defense, and they don't really know exactly how good any particular strategy is until it's tested in court. One good thing about this approach is that very few patent reviewers will object to a claim based on something written elsewhere in the specification, but it's pretty hard to get around the generality pulled in by that ITBAT. Then again, do you really want to be the guinea pig? Avoid means-plus-function language if you at all can.

Beware the claim language in your example patents may be out of date.

This is a good example also of the danger in using your example patents as your sole reference to how to structure your claims. Unless those patents are very recent, they can lead you astray of newly laid tripwires such as this.

9.4.7 Do Not Use Trademarks

Some words sound like trademarks because they once were...

while other seemingly generic terms will surprise you.

In keeping with your objective to maximize generality, never use a trademark in your CLAIMS. Like the rest of us, you probably consume way too much "cola" not "Coke™". You treat those owees with an "adhesive bandage", not a "Band-Aid™"…and ice skaters hope to hit the ice right after its been smoothed by an "ice resurfacer", not a "Zamboni™". These are such great products that we generally think of their trademarks as synonymous with the generic name for the product, but they are not. Many trademarks do gradually become the colloquial generic term for something, and you may not even be aware that many words in common use actually once were, or currently are, trademarks. I say "once were" because if the owner of a trademark does not very actively protect its use and constantly reassert its trademark status, a court can rule the term no longer a trademark. Here are some words I pulled out of Wikipedia™ that you may be surprised to learn were once trademarks, but now are legally generic in the U.S.:

Cellophane	**Heroin**	**Mimeograph**	**Trampoline**
Crock Pot	**Kerosene**	**Pilates**	**Webster's Dictionary**
Dry Ice	**Cellophane**	**Thermos**	**Yo-Yo**
Escalator	**Lanolin Linoleum**	**Touch-tone**	**Zipper**

Many words in common use are still active trademarks.

Of course, there are many more, and the trouble is, it can be quite difficult to know what's a trademark and what's not. For instance, here's a list of some words in common use that are still classified as trademarks:

Trademark	Generic Name
Adrenalin™	epinephrine
Bubble Wrap™	air-filled plastic wrapper
Frisbee™	flying disc
Hacky Sack™	foot bag
Jacuzzi™	hot tub or whirlpool bath
Jell-O™	gelatin
Kleenex™	facial tissue
LEGO™	building bricks or building toys
Mace™	pepper spray or tear gas
Photoshop™	photo manipulation software
Ping Pong™	table tennis
Popsicle™	ice pop
Post-it™	sticky notes
Q-tips™	cotton swabs
REALTOR™	real estate broker
Rolodex™	rotary card file
Saran Wrap™	plastic wrap
Scotch Tape™	adhesive tape
Sheetrock™	drywall
SPAM™	canned luncheon meat
Styrofoam™	polystyrene thermal insulation
Vaseline™	petroleum jelly
Velcro™	hook-and-loop fastener
Xerox™	photocopy/photocopier
Zamboni™	ice resurfacer
Ziploc bag™	zipper-closeable bag

Frequency of use is not a good indicator of whether or not a trademark has become generic.

I expect that some of these come as no surprise to you, while some of them do. Just because a word has achieved common usage, doesn't immediately invalidate its trademark status, and, after all, how common do I mean when I say "common usage" anyway? As seems to be the case for everything these days, it's up to a court to decide; but a lot of it comes down to how actively the trademark owner continues to assert ownership, and so one cannot judge simply by its frequency of use. Also, some words retain their trademark status in other countries, even if they have been ruled generic in the U.S. Regarding words you intend to use in your CLAIMS where you are even remotely unsure of the trademark status, the rule is: "When in doubt, check it out". The internet is a good tool for this. I'd start by Googling™ (there's a trademark that's becoming a word) the word in question plus "trademark".

The internet is a good place to find out if a word in question is a trademark.

There are three reasons you should not use any trademarks, current or obsolete, in your CLAIMS:

- Trademarks are seldom the most generic term, even when their trademark status has officially been rescinded. They open a door for creative interpretation by challengers that you don't want.

  Trademarks are usually not fully general.

- You can't control how the definition of a trademark evolves over time – your claims could actually become obsolete. Coke™ is free to change their formula any time (as they have occasionally), and "Model A" actually referred to two different cars manufactured by Ford decades apart. It is not uncommon for a trademark to be reused to represent an entirely different product. Note also that whereas patents expire, trademarks do not. So, whereas initially a trademark may refer exclusively to a single product protected by a patent, the day that patent expires many equivalent products may hit the market, but under other trademarks or generic labeling. If you used the original trademark in any of your claims, depending on how you used it, those claims may also effectively expire.

  To what a trademark refers can change over time.

- You're actually hurting the trademark owner by using it, so you may incur third party intervention. The fact that the Otis Elevator Company had used the word "escalator" in later patents actually was brought as evidence against them when Haughton Elevator Co. challenged the trademark in 1950. If a trademark owner gets wind of your using their trademark in the place of a generic in your claims, they can raise objections to the patent examiner (and must in order to look after their own interests).

  The trademark owner may object to your using it.

9.4.8 Do Not Claim the Result

You can claim a means to an end, but not the end itself.

Whereas the law specifically allows you to describe the element of an apparatus by its function, you cannot claim the function itself. This common mistake is referred to as "claiming the result". Consider the claim:

3. A belt that automatically changes length to maintain constant pressure on the wearer.

This is not a valid claim, because it claims the function of changing length to maintaining constant pressure, but not the means to do so. This should read something more along the lines of:

3. A belt comprising:

   a) a strap;

   b) a buckle; and

   c) a constant tension spring.

Claim elements may only be steps or things.

In the revised version, the claim contains elements that are exclusively things. If it became necessary to mention intent, say because it is necessary to limit the application to avoid infringement on prior art, we could write:

3. A belt comprising:

   a) a strap;

   b) a buckle; and

   c) a constant tension spring such that its length automatically adjusts to maintain constant pressure on the wearer.

You can include the result in addition to the means, but remember that will narrow your claim.

Here we have not claimed the result, but used it to qualify the apparatus comprised of the listed elements (but it really is best to find a way to leave the result out of it entirely). Generally as long as the result is introduced after the list of elements with "such that", "wherein", etc., you'll be OK.

9.5 QUICK QUIZ

As you continue to read through examples in this text, scrutinizing their quality should become a reflex.

What else is wrong with the above claim? Well actually, at least a couple of things. If you're catching on, the adjective "constant tension" should immediately have grabbed your attention. Do I really need that in my most general claim? No. Why would I let someone else copy my invention with a non-constant tension spring? It won't work as well, but sitting in a wrapper on a store shelf, it wouldn't look any different. Of course, the remedy is to first call out the invention omitting the adjective, and then bring it in later in a dependent claim.

What else is wrong? Could I find a more general term than "buckle" to serve the function of latching the belt? Once upon a time, I'd just use means-plus-function language, but we both now know that that's been sabotaged. I'd really have to think that through, though. In the end, I'd probably write the claim so as to exclude the buckle from the base claim entirely, and then put it in a dependent claim (I'd still have to think of a more general term, or use means-plus-function language and then comprehensively list all of the options in the DETAILED DESCRIPTION OF THE INVENTION, plus a nice juicy ITBAT).

9.6 OTHER COMMON MISTAKES

There are many mistakes besides bad language choices that unnecessarily narrow claims.

I've already discussed how you should avoid adjectives because they narrow the claim. Remember, everything in your claims constrains them. By now, you should have gathered that there are many equally effective ways to inadvertently narrow your claims other than bad language choices. It will help to examine a couple of common mistakes below.

9.6.1 Avoid Including Unnecessary Supporting Elements

In your most general independent claim you should distill your invention down to the minimum elements required to define it.

Be careful not to add any extraneous elements to your CLAIMS that do not really comprise part of the invention. In my example of the belt on the preceding pages, I was tempted to let my search for a general term for a buckle distract me from the more important fact that I may not need that element at all in my most general claim. If I have a clever way to spring load a belt, is the buckle even part of the invention? Not unless the spring's incorporated into the buckle somehow.

Write multiple claims with varied strategies – for each element of IP, at least one, but only one, needs to be successful.

When you find a flaw in a claim, keep looking – most have more than one weakness.

As a matter of fact, do I even need to call out the strap? I want my independent claim to be as general as possible, so I need to avoid confusing extraneous paraphernalia associated with using my invention with the invention itself. What I'd really do for this belt concept is try to call out only the spring mechanism. I'd then call out its integration with the strap in a dependent claim. Clearly, the way the notional claim is written, I'd need to add some more details specific to the spring mechanism (which the example just called a "constant tension spring"). You could also both call out the spring mechanism separately and have a claim very much like the example for an integrated assembly. Remember your claims work independently from each other (except, of course, for their parents), so your CLAIMS should take a shotgun approach – claims that don't turn out to be useful won't do any harm; you just need to make every possible attempt to get one that really will hold water.

Here's another example. Consider the invention of a special progressively geared crank for a vice that automatically adjusts the gear ratio as output torque increases. I could be tempted to include the vice as an element of the claim, but I'd be better off just claiming the crank assembly in my most general claim. In a dependent claim, I could then claim a vice using the crank assembly.

9.6.2 Don't Get Stuck on Your Preferred Embodiment

It can be difficult, but absolutely vital to think beyond the specific details of the example embodiment in your specification or your prototype.

Failing to see past the preferred embodiment is also a very common problem, leading to unnecessarily restrictive patent claims. I've already given a good example in our constant tension belt, where the natural preferred embodiment used a buckle as a fastener, but then realized I needed something more general in the claim. Are those really bearings, or can my part simply rotate about something else, or just be rotatable. Do I need hinges or can my parts just be pivotally connected? Do those even need to be wheels? Do those fasteners have to be screws? (Gotcha! You should have balked at that last one. I can't even imagine mentioning specific fasteners in a claim unless I'm actually patenting a type of fastener. If something is fastened to something else, that's exactly as much detail as I'd include, and even then I'd be thinking about how I could be more direct about the functional interaction.)

You may be coming back to just how useful means-plus-function language was for avoiding this. Indeed it was, but the rules are, these days, for better or for worse, that you're simply going to have to work hard to think up more original general terms. At least the body of prior art is chock full of examples if you get stuck.

9.7 Method vs. Apparatus Claims

Methods are also patentable.

Many apparatus inventions are also accompanied by a method invention.

Statistically speaking, your invention is probably a device or apparatus, but keep in mind that methods of doing things are also patentable (provided they are novel, useful, and non-obvious). Even if you think of your invention as a device, when you write your claims, make sure to consider if the way your product is used represents a new method. Whereas the elements of apparatus claims are things, the elements of a method claim are steps – simply claiming the use of your invention won't fly as a method claim. Take a look at the example claims below for a hat with a built-in sponge for keeping the wearer cool when dipped in water:

> 1. A hat comprising:
>
>   a) a brim; and
>
>   b) an absorbent element.
>
> 2. A method of using the hat of claim 1. ☒ **No**
>
> 3. A method of using the hat of claim 1, comprising the steps of: ☒ **No**
>
>   a) wetting the absorbent element; and
>
>   b) placing the hat on a wearer's head.
>
> 4. A method of cooling a subject's head, comprising the steps of:
>
>   a) wetting an absorbent element attached to a hat; and
>
>   b) placing the hat on a wearer's head.

If the order of steps in a method claim is arbitrary, write it that way in your CLAIMS.

The example is simplistic for brevity, but should suffice. Claim 2 won't fly (because you can't simply claim using your invention as a novel method), but claim 3 is correctly structured in that it calls out specific steps as elements of the method (but there is also a problem with claim 3, to which we will come shortly). If use of your invention involves a novel method, there should be a number of elemental steps that you can define and outline. Note that the order of the elements is arbitrary – and you should always avoid words that restrict the steps to be performed in any particular sequence unless they truly must be.

As should be expected, maximum generality is just as important in method claims as apparatus claims.

As with apparatus claims, always concentrate on being as general as possible. Note, for instance, in element a) of claim 3 I wrote "wetting the absorbent element" instead of "wetting the absorbent element with water". Water would be the most obvious fluid to use with our hat that cools the wearer's head, but we simply don't need the arbitrary limitation. If we wrote "with water", someone might try to get away with selling the same hat simply by including instructions to use a different fluid (but knowing the end user would still just use water).

Many examiners will object to "mixed claims".

Claim 3 above does have one open issue, however. It is what referred to as a mixed claim, because it references claim 1, which addresses an apparatus (whereas it is a method claim). Many patent examiners will not allow mixed claims. Also, note that as of November 1, 2007 claims of this type count as independent claims, even though it has another claim as an antecedent. This is because it is a method claim, and the antecedent is an apparatus claim. It would only be interpreted as a dependent claim if its antecedent was also a method claim. In general, you should avoid mixed claims – if the method is truly novel, you won't really need to cross-reference your apparatus claims, e.g. claim 4 above. Method claims also commonly reserve a (novel, useful, and non-obvious) method of manufacture of an invention.

You cannot mix apparatus and method claim elements.

Make sure you are clear on the distinction between method and apparatus claims, and their elements. What you absolutely cannot do is mix elements of different types within a single claim, e.g.:

---

5. A hat comprising:

    a) a brim;

    b) an absorbent element;   ⊠ No

    c) wetting the absorbent element; and

    b) placing the brim on the crown of a wearers head.

Here elements a) and b) are apparatus elements, but c) and d) are steps, so claim 5 above doesn't make any sense. A claim may either be for a method or an apparatus, but not both.

9.8 TEST YOURSELF

See if you can spot what's wrong with the following claims. Each claim has at least one error, and some have more than one. I recommend jotting your responses on a piece of scratch paper before you check out the answers on the pages that follow.

---

1. A gun comprising:

    a) a source of antimatter fuel;

    b) an antimatter emitter;

    c) a focusing element.

2. The gun of claim 1 where the fuel source is a reservoir of anti-protons or anti-hydrogen.

3. The gun of claim 2 where the reservoir is a magnetically-confined detachable container.

4. The gun of claim 1 without the focusing element thereby affecting a wide-angle beam.

5. A projection table comprising:

   a) a horizontal surface; and

   b) a projector positioned to project an image onto the table.

6. The projection table of claim 5 where the likeness of a ping pong table is the projected image.

7. A musical instrument comprised of:

   a) a rotatable spindle;

   b) One or a multiplicity of glass ridges affixed to the rotatable spindle.

8. A method of producing tones comprising:

   a) the musical instrument of claim 7

   b) rotating the spindle and the affixed ridges

   c) applying one or more operator's fingers to the ridges affixed to the rotatable spindle.

9. The musical instrument of claim 7 where the rims are affixed to the rotatable spindle by epoxy adhesive.

10. A fuel delivery system comprising:

    a) a fuel

    b) an injector;

    c) a pump;

    d) a tank; and

    e) a blanket fluid that is lighter than, and immiscible with, the fuel.

**Jot down your responses before you look
at the answers on the following pages.**

Answers:

1. A gun comprising:

    a) a source of antimatter fuel;

    b) an antimatter emitter;

    c) a focusing element

    The "and" is missing after the semicolon following "emitter"

2. The gun of claim 1 where the fuel source as a reservoir of anti-protons or anti-hydrogen.

    The claim should not use "anti-protons" and "anti-hydrogen" in the alternative. This claim should be broken into two separate claims, one for "anti-protons", and one for "anti-hydrogen".

3. The gun of claim 2 where the reservoir is a magnetically-confined detachable container.

    Do I need the adjective "magnetically-confined"? to describe the detachable container? Do I even need "container" to bring in the detachable concept? No. The claim would better read "...where the reservoir is detachable".

4. The gun of claim 1 that achieves a wide angle beam by removal of the focusing element.

    This claims the result "achieves a wide angle beam", which is a no-no. It also does so by removing element c) from claim 1, which a dependent claim can't do. To incorporate the intent, a claim would have to be added ahead of claim 1 that omitted element c), and then the element reintroduced in a dependent claim.

5. A projection table comprising:

    a) a horizontal surface; and

    b) a projector positioned to project an image onto the table.

    There are actually two things wrong with the claim. Firstly, the terminology is inconsistent. Element b) should have read "...onto the surface." Secondly, the word "horizontal" is absolute. A table that was tilted by one degree would be substantially equivalent, but not protected. We need something more general – if we couldn't be more creative, we might just write "a surface oriented within twenty degrees of horizontal".

6. The projection table of claim 5 where the likeness of a ping pong table is the projected image.

    Ping Pong™ is an active registered trademark of Parker Brothers. The correct generic would be "table tennis".

7. A musical instrument comprised of:
   a) a rotatable spindle; and
   b) one or a multiplicity of glass ridges affixed to the rotatable spindle.

Two things are amiss in this claim. Firstly, the claim commits the cardinal sin of using "comprised of" instead of "comprising". Secondly, the adjective glass should be omitted here and brought back in a dependent claim.

Incidentally, this claim describes Benjamin Franklin's personal favorite among his inventions, which worked according to the same principle by which glasses emit an audible tone when you rub the rim with a wet finger. Franklin patented none of his inventions, but freely released them into the public domain, stating that he already had enough money. Admittedly, by the time he retired from his nationwide printing/publishing chain at age 42, he had amassed quite a bit.

8. A method of producing tones comprising:
   a) the musical instrument of claim 7;
   b) rotating the spindle and the affixed ridges
   c) applying one or more operator's fingers to the ridges affixed to the rotatable spindle.

This is a mixed method/apparatus claim, since claim 7 describes an apparatus and claim 8 describes a method. Hence, a) is an apparatus element whereas b) and c) are steps of a method. This could correctly have been written:

   8. A method of producing tones using the instrument of claim 7 by:
      a) rotating the spindle and the affixed ridges
      b) applying one or more operator's fingers to the ridges affixed to the rotatable spindle

Nevertheless, it would be better still to describe the method without referencing claim 7, and since the parent claim is an apparatus and this is a method claim, it will still count as an independent claim, regardless.

9. The musical instrument of claim 7 where the rims are affixed to the rotatable spindle by epoxy adhesive.

The terminology is inconsistent. In this claim we are now referring to as "rims" what we previously identified as "ridges" in the parent claim.

Also, assuming there was something important to manufacture about gluing the assembly together that might block others from being able to make cost-competitive rip-off products, why specify "epoxy"? Clearly we can just write "... by an adhesive".

> 10. A fuel delivery system comprising:
>   a) a fuel
>   b) an injector;
>   c) a pump;
>   d) a tank; and
>   e) a flame retardant blanket fluid which is lighter than, and immiscible with, the fuel.
>
> *There's too much extraneous stuff here. Admittedly, this may have been hard to catch, since this is just a made-up example and you have no familiarity with the details of the invention, but a laundry list of conventional elements like this should give you pause. If I'm patenting the concept of a tank where an inert fluid sits on top of the fuel (to blanket it and prevent fire if the tank ruptures), why bring the injector and (fuel) pump into it?*
>
> *Look at the clause "immiscible with". I've included it (and "lighter than") in the scenario anticipating that I will need to be more specific, because the fact that the blanket fluid has both these qualities is fundamental to the concept. But "immiscible with" is an absolute. No two fluids are 100% immiscible, so, more correctly, we should say "has low miscibility with" or something like that.*
>
> *Notice that the adjective "flame retardant" didn't come into the blanket fluid description. I'd hold off and bring that into a dependent claim.*

Most will find it easier to write their own claims than review others'.

So how did you score? Be sure, of course, to go back and review the relevant section(s) if you missed something. You may well have found the exercise tedious and difficult, which is normal. The upside is that I believe you will find it more straightforward to avoid pitfalls in your own claims than nit-pick them out of claims someone else wrote.

9.9 CLAIMS BY ORGANIZED DESIGN

Making a table of potential elements can be a useful tool for constructing your claims.

You should now be in pretty good shape to tackle your CLAIMS. One method to organize your thoughts is to create a tabular "claims design matrix" like the one shown in Fig. 9-1. Listed in the left column are the building blocks of your claims – elements and packages of qualifying language. Each column to the right represents a claim. As a general approach, one may start simply by listing claim elements you plan to use on the left, and then placing "X's" to denote which elements comprise which claims. So doing will provide you with a bird's eye view of the complete scope of all of your claims, allowing you to sculpt the entire field simultaneously so as to optimize your coverage. A second step (although you will generally iterate back and forth between these steps) will be to figure out which claims should be dependent claims, and of what parents, denoted in the row along the bottom. As I do this, I like to change the "X's" that are inherited from their parents to "O's". For multiple dependent claims (if you really must), use multiple columns labeled with the same number (plus a letter to distinguish them, if you find it helpful), one for each reference to a parent claim. (Remember that when you patch the language together into a claim, you must back-reference in the alternative, not conjunctive.)

Elements & Qualifying Epilogues	Claims																							
	1	2	3	4	5a	5b	6	7	8	9	10	11	12	13	14	15	16	17	18	19	20	21	22	23	24
A pastry	X	O	O	O	O	O																			
A stick providing a means of holding the pastry	X	O	O	O	O	O																			
A removable wrapper surrounding the pastry.	X	O	O	O	O	O																			
...where the pastry is circular, having a hole in the center		X																							
...where the pastry is a doughnut			X				O																		
...where the pastry is at least partially frosted.				X			O																		
... with sprinkles					X	X																			
Parent Claim:		1	1	1	3	4																			

Fig. 9-1 CLAIMS design matrix for the "Doughnut on a Stick"

You can start the claim elements list with shorthand notes if you prefer, but right in that leftmost column is where you should craft the wording of each element with surgical precision. This is the last place in your patent application where you want to be in a hurry! Scrutinize every word of those claim elements. It'll be easier and save you time to do it now, rather than after you've transcribed the table into a set of claims.

You'll be more consistent if you refine the specific language of each clause of your claims directly in the CLAIMS Worksheet.

Once you've completed the table the hard work is done – the task of converting the worksheet to claims is purely clerical. For each claim, cut and paste the elements; seam them together under your heading, e.g. "A food product comprising:"; format; and presto! Your CLAIMS are finished.

Action Step 18

Use the worksheet provided in Section 6 of the Application Workbook to create your claims.

When finished, transcribe them into outline format under "I claim:" on the CLAIMS pages of Section 3 of the Application Workbook.

9.10 A FINAL WORD OF ADVICE

> **Have a competent patent attorney check out your claims.**

Having a patent attorney check your work will cost a small fraction of having it written from scratch.

Although hopefully you have fully internalized every word of this section, the fact remains that it is exceedingly difficult to be certain that your CLAIMS conform to, and take full advantage of, the latest fashion in interpretation of the law. As such, I strongly recommend that you have an experienced patent attorney review your patent application, and above all, your CLAIMS. Of course, that won't be free (unless you have a buddy who happens to be a patent attorney), but since you've already done almost all of the work, the cost should be very moderate. Don't let anyone sell you a patent search – you already did a very thorough one back in Chapter 3 (right?).

Good attorneys are, indeed, hard to find.

How do you find a patent attorney? Well, that's easy. Unfortunately, what you really need is an attorney who is both honest (somewhat difficult to find, depending on your standards) and has considerable experience in the field of your invention (very difficult to find, unless your invention relates to toilets – as I mentioned previously, toilet inventions seem to be very common).

Integrity is indispensable to your interests.

Verifying integrity is something you should do with anyone whom you propose to hire for anything. I can't say that I have worked with so many attorneys that I can give you a meaningful statistic regarding what fraction will or won't rip you off, but I expect it's about the same proportion as with most contractors, which means watch out. There are two simple steps that you can take to greatly decrease your chances of being taken in by a poser, scammer, drunkard, whatever:

1 **Get a Quote.**

Feel free to show them your work and request a quote. Be very clear that all you want them to do is review and edit a patent specification you've already written. An honest attorney will understand that you're just a regular Joe who still has to pay the rent, and should be willing to provide you an estimate. (He/she may want to see the document first, or may just ask you how many pages and how many claims you have.) There's nothing wrong with being a normal citizen on a budget. If you sense even the slightest hint of contempt (but you probably won't), get out while the getting's still good. Always get a minimum of three quotes and preferably five. Of course, don't necessarily go with the lowest bidder, but certainly don't go with the one that's three times higher than the mean.

2 Ask for References.

What you need is to talk with people who have worked with that particular attorney in the past. If you don't know of the attorney through someone who can give you the straight dope, ask for references. This isn't trial law (where confidentiality rules. Beware what shrinks from the light of day!) – a good attorney should be able to provide some. A second good indicator is openness. Firms that woo your business by providing a lot of free information have a winning strategy, in my opinion (say, on a website such as Brown & Michaels, http://www.bpmlegal.com/index.html (for the record, I don't receive kickbacks from them and have not had the occasion to work with them, but their site has been very helpful to me)).

Finding an attorney who is an expert in the field of your invention may be difficult.

Licensing often has the advantage that the licensee's expert attorneys take over prosecution of the patent.

You don't need the attorney to be local.

Courtroom experience is important.

Finding an attorney who is an expert in your field of invention may be quite a bit more difficult. Who, else, however, can better review your CLAIMS? In this respect, many of the best patent attorneys unfortunately aren't accessible to you. They work for firms that cater to the big dogs. It's not personal; who would you rather represent: Mongoose™ or Joe-Bob's Bikes? Larger companies simply provide a much more steady supply of work and have bigger budgets for each project. But certainly, if you represented Mongoose™, what an expert on bicycle-related patents you would become! This is one place those planning to license their patent have a distinct advantage. As previously discussed, if you license your invention during the one-and-a-half to two years between when you file and the arrival of the first office action, often the licensee will take over prosecution of the patent (exactly because it's in their best interest to make sure the patent is solid, and they have attorneys who are experts in the field of invention).

On your own, you simply may not be able to recruit an expert in your field, but you should make your best effort to do so. Remember, you don't have to shop near home. Since you've already completed the specification, you really don't need a lot of face-to-face time with the attorney (which costs money – we're trying to avoid that). You can do all of your initial fact-finding by correspondence, and once you've selected a firm, all they need is the document. If you really can't find someone familiar with the field of your invention (meaning they've worked at least several patents in the field), a generalist will still at least be able to provide important feedback regarding the basic structure and strategy of your CLAIMS, and that's better than nothing.

Another thing you want your attorney to have is trial experience. Battling it out in court is how they really learn what works and what doesn't. A patent attorney who hasn't gone to trial is still just a greenhorn, and you want a veteran.

Once you receive an edited draft of your specification, don't assume that just because a patent attorney did it that it's good-to-go. You must

Check the work –

A patent attorney may know the law, but you know your invention.

review the work. Hopefully your attorney knows the law (but remember, half of them are in the bottom 50%), but he/she won't necessarily immediately grasp the complete breadth of your invention. It's not uncommon for an attorney to do an excellent job covering the most prominent features of your invention, but leave out something a little more subtle that you know as the inventor to be quite important. If your CLAIMS come back with something you specifically intended to claim deleted, question it.

A patent attorney should be willing to sign an NDA.

Lastly, unlike a potential licensee (for reasons we will come to in Chapter 11, a patent attorney will have no reason to object to signing a non-disclosure agreement (NDA), and should be able to provide you a standard form.

When interviewing a prospective patent attorney, always:

- **Get a quote – Make sure to be clear that all you want them to do is review/edit an already drafted specification**

- **Ask for references.**

- **Ask about experience in the field of invention.**

- **Ask about trial experience.**

- **Ask if they will sign an NDA before you send them your specification.**

10 Finishing Up

10.1 ABSTRACT OF THE DISCLOSURE

The ABASTRACT OF THE DISCLOSURE should allow quick assessment of the field and general nature of your invention.

One small task now remains to complete the text of your application, and that is to write the ABSTRACT OF THE DISCLOSURE. This abstract provides a several sentence paragraph describing the invention to serve as a tool for USPTO personnel and patent searchers to understand the field and nature of your invention, particularly making note of what is novel about your invention. Limited to a mere 150 words, this is the easiest task of all.

Action Step 19 Type your ABSTRACT OF THE DISCLOSURE under the heading provided in the Application Workbook.

10.2 Completing Your Patent Application

You now only have a few forms to fill out.

So, with the completion of your specification, drawings, and abstract, the lion's share of the work is now done, and you now need only fill out a few forms to complete your patent application. The package you send the USPTO should include, in the order listed, the following:

1. **A Utility Patent Application Transmittal Form**
2. **A Fee Transmittal Form and appropriate fees**
3. **Any Application Data Sheets you wish to file (optional)**
4. **The elements you completed in the preceding sections:**
    a. **Your specification**
    b. **Your CLAIMS**
    c. **Your ABSTRACT OF THE DISCLOSURE**
    d. **Your drawings (unless your application doesn't need them)**
5. **A Declaration (or Oath)**
6. **A nucleotide and/or amino acid sequence listing (if necessary)**

We will address these one-by-one. Note that the USPTO has made all of its forms available in Adobe Acrobat™ (PDF) format via the internet. Here's a link to the main forms directory:

http://www.uspto.gov/web/forms/index.html#patent

Most government document references are actually only an internet search away.

The patent filing forms are really quite simple and fairly self-explanatory. There are line items here or there that might raise questions, however; so, for the sake of thoroughness, you'll find guidance for each one you will need in this section. More information is really easy to get, if you have a reason to be interested. Much of the time, when inventors get hung up in the applications process, it's because they see some reference such as "per 13 CFR Part 121", which naturally they know nothing about. Unfortunately, few people realize how easy this information is to get. Can I tell you where to find a listing of all those CFR's? I could if I bothered to find out, but I don't need to, and neither do you. If you want to see the text of 13 CFR Part 121, just open up any good search engine, cut and paste the reference into the search box, hit return, and it should come right up. The language of most of these documents really isn't as convoluted as you might expect, either. So, you're far from as helpless as that intimidating reference might first make you feel.

You can complete the USPTO's forms electronically, but you can't save the filled-out copy.

These forms can be completed electronically before printing them out, but you cannot save the completed form to your hard drive (I have no idea why the USPTO set it up that way), so you should print them out as soon as you complete them. Fortunately, the forms are quite short, so if you want to change something, it's pretty fast to just retype the information using your hardcopy for quick reference. You'll want to keep a copy for yourself, so print two.

10.2.1 Utility Patent Application Transmittal Form PTO/SB/05

Form PTO/SB/05 is basically a cover sheet listing the contents of your application.

For conventional filing by mail, every application must have a cover form listing the complete contents of your submission (it's really simpler to just stick with the form, but technically you may also use a letter (with no real advantage) if you're into doing things the hard way). Think of this like a fax cover sheet, its intent being to provide a reference so that the patent examiner will know if something you intended to include is missing. With the introduction, a number of years ago, of the USPTO's EFS-Web electronic filing option (recommended, see Section 10.3) with entry fields that are essentially redundant with the transmittal form, there exists a common question regarding whether or not PTO/SB/05 remains necessary, to which the USPTO responds on its EFS-Web FAQs page as follows:

> *"When filing a new application by EFS-Web, a signed transmittal form or a signed application data sheet (ADS) is recommended for identification purposes. However, a signature is not required to obtain a filing date for a new patent application."*

Designate Form PTO/SB/05 as a "Miscellaneous Incoming Letter" when uploading on EFS-Web.

Thus, while the USPTO makes a clear indication that they will not insist upon it, Form PTO/SB/05 is both recommended as best practice and very simple to complete, and so a good idea. Note also that, although the USPTO does not require you to sign the form, since you must still print, sign, and scan a PTO/SB/01 Declaration Form (Section 10.2.5), you may as well do them both together. Note, however, that PTO/SB/05 is not listed as an option in the EFS-Web drop-down menus on the "Attach Documents" page. I have verified with the USPTO that the correct identifier to select is "Miscellaneous Incoming Letter" under the "General Transmittal" category.

Completion of Transmittal Form PTO/SB/05 begins with completion of "First Inventor" and "Title" fields at the top. Moving on to the center box, you will find the following additional fields:

1. Fee Transmittal Form	This form (PTO/SB/17) is where you report your calculated filing fees. You will be paying them with the submittal of your application, so check the box.
2. Applicant claims small entity status.	If you are a small entity, you get 50% off on all fees. Basically, if you are an independent inventor (or inventors), or qualify as a small business, and have not licensed or sold the patent to anyone, other than the government, who doesn't also count as a small entity, you can check the box. See Section 10.2.2 below for a little more on who and what counts as a small entity.
3. Specification	This is the specification you've worked so hard to complete, so check this box.
4. Drawing(s)	Obviously, if you have drawings you need to check this box.
5. Oath or Declaration	This is another mandatory form you will be completing, so check this box.
6. Application Data Sheet. See 37 CFR 1.76	On initial filing, a form PTO/SB/14 Application Data Sheet would be redundant with the other forms. This form can be used later if information such as the correspondence address needs to be updated. There is one special field pertaining to secrecy orders (for patents pertaining to classified materials). I'm thinking that if this applies to you, you already have people taking care of this for you. The USPTO has a link to a pretty thorough set of instructions regarding this form in their forms directory.
7. CD-ROM or CD-R	This refers to the inclusion of a CD-ROM for the special cases where applicable, such as computer programs.
8. Nucleotide and/or Amino Acid Sequence Submission	If you're a molecular biologist, you'll know what this is. It can also be included on a CD-ROM.
9. Assignment Papers (cover sheet (PTO-1595) & document(s)) *and* 10. 37 CFR 3.73 Statement Power of Attorney (when there is an assignee)	If you've signed an exclusive licensing agreement (that will generally involve assignment of the patent to that entity), you'll need these; however, your licensee's patent attorneys will also take over prosecuting the patent, so you'll no longer need to worry about managing the details of what forms need to be filed. You also will no longer qualify as a small entity, but your licensee will be taking care of the fees.
11. English Translation Document	Necessary only if you typed your original patent in a language other than English and then had someone else translate it for you, otherwise leave this box blank.

12. Information Disclosure Statement (PTO/SB/08 or PTO-1449) Copies of foreign patent documents, publications, & other information	This is a form for forwarding references to other patents relevant to your own to the patent examiner. You won't need this for filing your application. The reference lists you see in published patents are compiled by the patent examiner, not the applicant. Generally it's not in your interest to point out similarities between your patent and prior art.	
13. Preliminary Amendment	You won't need this for your initially filing. Generally you'd file a preliminary amendment if you wanted to change something in your application (say adjust the claims based on something you learned after the initial filing) before you received the first office action back from the USPTO.	
14. Return Receipt Postcard (MPEP 503)	This is a postcard that you can provide for the USPTO to stamp and mail back to you to verify the date of receipt of your patent application. You should include this only if you are not filing electronically (more details later).	
15. Certified Copy of Priority Document(s)	Include these if you are referencing foreign patent filings to establish a priority date. Obviously, most applicants won't be checking this box.	
16. Nonpublication Request under 35 U.S.C. 122(b)(2)(B)(i).	Generally, your application will become publicly viewable 18 months after filing. You can request, via form PTO/SB/35 to have your patent application not be published until awarded, but there's a hitch. If you file a patent application for the same invention in another country that would cause it to be published within 18 months, you must notify the USPTO within 45 days of having done so or they will abandon your application. I generally wouldn't bother with this.	
17. Other:	Pretty clearly, anything you include that doesn't fall under one of the listed categories gets indicated here. It's unlikely that you will ever need to send anything qualifying as "other" to the USPTO in association with your patent application.	

"Continuations" are patent applications that modify or add to an existing patent.

Moving down the page, in Section 18 you'd indicate your application to be a continuation (to a previous patent) if that were the case. Continuations can qualify content of your original patent (without changing the date of priority) or incorporate additional intellectual property into your patent (carrying the continuation filing as its priority date). The specific logistics of filing continuations is outside the scope of this guide, but it's not much different than filing the original application.

At the bottom, specify to what address you wish the USPTO to send any correspondence related to your application. Sign and date, and you're ready for the next form.

> **Action Step 20**
>
> Go to the USPTO's forms website, select PTO/SB/05, and fill it out.
>
> http://www.uspto.gov/web/forms/index.html#patent
>
> Print to PDF. If filing by mail, print two copies when finished. (Remember, you can't save it electronically in an editable format.)

10.2.2 Fee Transmittal Form PTO/SB/17 and Appropriate Fees

Your application will not be processed until the correct filing fees are paid in full.

You'll recall that I previously mentioned that the USPTO is entirely self-supporting, with its primary source of income being patent fees. That's good news to us taxpayers, but it also means they have to take fees quite seriously. One of the first things they will do is check your fee calculations – if you've not computed your fees correctly (and provided payment in your filing package), they will notify you and your application simply will not be processed until you do.

Small Entity Status

The independent inventor qualifies for reduced fees as a "small entity".

Above I touched on the fact that if you qualify for small entity status (and haven't already licensed your invention to any disqualifying entity), you get a 50% discount on filing, issue, and maintenance fees. Ninety-nine percent of the users of this guide can expect to qualify as small entities. Here is an excerpt from MPEP 509.02 which you may find useful:

(a) *Definition of small entities*. A small entity as used in this chapter means any party (person, small business concern, or nonprofit organization) under paragraphs (a)(1) through (a)(3) of this section.

   (1) *Person*. A person...means any inventor or other individual (*e.g.*, an individual to whom an inventor has transferred some rights in the invention) who has not assigned, granted, conveyed, or licensed, and is under no obligation under contract or law to assign, grant, convey, or license, any rights in the invention. An inventor or other individual who has transferred some rights in the invention to one or more parties, or is under an obligation to transfer some rights in the invention to one or more parties, can also qualify for small entity status if all the parties who have had rights in the invention transferred to them also qualify for small entity status either as a person, small business concern, or nonprofit organization under this section.

   (2) *Small business concern*. A small business concern...means any business concern that:

      (i) Has not assigned, granted, conveyed, or licensed, and is under no obligation under contract or law to assign, grant, convey, or license, any rights in the invention to any person, concern, or organization which would not qualify for small entity status as a person, small business concern, or nonprofit organization; and

(ii) Meets the size standards set forth in 13 CFR 121.801 through 121.805 to be eligible for reduced patent fees. Questions related to standards for a small business concern may be directed to: Small Business Administration, Size Standards Staff, 409 Third Street, SW., Washington, DC 20416.

(3) *Nonprofit Organization.* A nonprofit organization, as used in paragraph (c) of this section, means any nonprofit organization that:

(i) Has not assigned, granted, conveyed, or licensed, and is under no obligation under contract or law to assign, grant, convey, or license, any rights in the invention to any person, concern, or organization which would not qualify as a person, small business concern, or a nonprofit organization; and

(ii) Is either:

(A) A university or other institution of higher education located in any country;

(B) An organization of the type described in section 501(c)(3) of the Internal Revenue Code of 19 86 (26 U.S.C. 501(c)(3)) and exempt from taxation under section 501(a) of the Internal Revenue Code (26 U.S.C. 501(a));

(C) Any nonprofit scientific or educational organization qualified under a nonprofit organization statute of a state of this country (35 U.S.C. 201(i)); or

(D) Any nonprofit organization located in a foreign country which would qualify as a nonprofit organization under paragraphs (a)(3)(ii)(B) of this section or (a)(3)(ii)(C) of this section if it were located in this country.

(4) *License to a Federal agency.* (i) For persons under paragraph (a)(1) of this section, a license to the Government resulting from a rights determination under Executive Order 10096 does not constitute a license so as to prohibit claiming small entity status.

The Leahy-Smith America Invents Act defines a third "Micro-Entity" status, under which any small entity who is named on fewer than five prior patent applications and made less than three times the median income in the year preceding filing is entitled to 75% off most fees. Unfortunately, enaction of this benefit has been delayed, probably at least until 2013.

Fees

Be aware that other fees come after the filing fee.

Notice above that I said small entities pay their fees, not fee, at 50%. The filing fee is not the only fee you will need to pay, just the first one. While the other fees are not really within the scope of this guide, I think it's only fair to mention them now, so that you know what you're on the hook for later. Your small entity utility patent filing fees (sent along with your application) will total $625, plus $30 for every claim over twenty, $125 for every independent claim above three, and another $225 for every multiple dependent claim (I told you they weren't cheap). Note that you will owe the USPTO another $300 when your application is published eighteen months after filing (no small entity reduction for this one), and $870 at the end of the application process, once the examiner is satisfied that your claims are valid.

You will not be granted a patent until these fees are paid. That may seem a lot, but not if you check out what it costs in other countries.

Once you have a patent, there are periodic maintenance fees which you must pay to keep your right to exclude others from practicing your patent, i.e., if you don't pay these fees, your patent goes into the public domain. Here's a schedule for maintenance intervals and fees:

You must pay periodic maintenance fees for your patent to remain active.

Patent Fees After the Initial Filing (Small Entity)

Publication Fee:	$300.00
Issue Fee:	$870.00
Due at 3.5 years:	$565.00
Due at 7.5 years:	$1,425.00
Due at 11.5 years:	$2,365.00
Surcharge - 3.5 year - Late payment within 6 months:	$75.00
Surcharge - 7.5 year - Late payment within 6 months:	$75.00
Surcharge - 11.5 year - Late payment within 6 months:	$75.00

There's a late fee, but at least you're not completely out of luck if you miss a deadline.

I've listed fees here to provide you a quick understanding of the approximate cost. Naturally, fees are subject to change without notice, so pay attention to the fees listed on the USPTO's fee computation form PTO/SB/17 (which we will address specifically in a moment). Notice also that there are fines for paying fees late. This is a good thing – it means you don't lose your patent just because you accidentally didn't file the right 27B|6 (pronounced "27B-stroke-6") on time (any Terry Gilliam fans out there?).

The filing fee doesn't come close to covering the actual cost of processing a patent application.

Now, your first reaction may be that for the USPTO to keep coming back to you for more money seems a bit like gouging, but consider – this is the real cost of processing, evaluating, publishing, and keeping track of patents; it's a lot of work. Paying someone, say, $50/hr, your filing fee would pay for about eleven hours of an examiner's time. (In reality, you don't just pay for the examiner's time, but all of the other support staff and miscellaneous overhead elements (like keeping the lights on) to which your fees would be applied in any contracting situation, so the net cost of examining your patent is probably more like $200/hr.) Clearly, it's going to take a lot more than one working day for the examiner to understand your invention, research the prior art, and scrutinize your claims, plus send you office actions and review your replies; so, it's not so much an issue of how much you pay, but when you pay. The fees you'll pay on your patent will total about $6,150. How'd you like to pay that all up front?

The fee schedule is all about taking care of small inventors. Keep in mind, at the corporate level where patents are entirely drafted and prosecuted by professionals, the filing fees are chump change. The USPTO has made

Patent fees are distributed over time to benefit the small independent inventor.

great efforts to make sure patents remain within reach of people just like you (after all, a nation's inventors are one of its greatest assets) by both reducing small time inventor fees by 50%, and also by not making you pay it all up front, which minimizes your risk. So, when your first file, you pay the non-refundable initial filing fee, but you're not further committed. If the process ends with your finding out you missed something in your patent search that invalidates the novelty of your invention, you don't owe any more money. At each of the maintenance intervals, you are free to re-evaluate the value of your patent. By the 3.5-year mark, you should have a pretty good idea if your invention is marketable. If it's not, you can just walk away without spending any more money (your patent won't cease to exist, but you will lose the right to exclude others from practicing). If it is, well then you're making money, and it'll be easy for you to justify paying the maintenance fee. The point is, you get to withhold most of the real expense until you know whether or not your patent is worth it, and that's a pretty smart business.

If your invention doesn't market well, you can walk away from the patent any time and stop paying fees.

Filing Out the Form

OK, speaking of business, let's get down to it. The USPTO provides form PTO/SB/17, where you can directly compute your fees based on the number and type of claims you have. You can find a schedule of all of the USPTO's patent fees here:

Filing fees are listed directly on the Fee Transmittal Form.

http://www.uspto.gov/web/offices/ac/qs/ope/fee092611.htm

You won't have to go hunting for filing fees, however. As I mentioned above, the USPTO provides them directly on the form, which is occasionally updated, typically at the beginning of the government fiscal year in October (so although updates are not annual, but rather fairly infrequent, just in case you should never use an old copy of the form – always download the latest revision). This is a very easy form:

Note that if you will be filing electronically (which you should – it costs $400 less, see Section 10.3), since Fee Transmittal Form PTO/SB/17 is built into the EFS-Web submission site you can skip Action Step 21. Is not necessary to also fill out the PDF version. As the web-based and conventional PDF fee computation forms are very similar, the following instructions are equally applicable for either.

---

1 Check the "Applicant claims small entity status" box on the upper left (unless, of course, it actually does not apply).

---

2 Fill in the filing date and the first inventor name in the box on the upper right.

3 Check the method of payment (PDF form only). For regular paper filing by mail I recommend a check, but you can use any major credit card if you are willing to fill out Credit Card Payment Form (PTO-2038). Check the "Charge fee(s) indicated below" box. You can also check the "Charge any additional fee(s) or underpayments of fee(s) under 37 CFR 1.16" and 1.17 "Credit any overpayments" boxes if paying by credit card and you're worried you'll make a mistake computing the fees (but it's pretty straightforward). Checks should be made out to "Director of the United States Patent and Trademark Office". For electronic submission you'll have a later opportunity to pay online by credit card.

4 Under "1. BASIC FILING, SEARCH, AND EXAMINATION FEES, you'll see there is an itemized list of "FILING FEES", "SEARCH FEES", and "EXAMINATION FEES". These all apply – you owe the total. These fees are itemized because they are attached to specific parts of the application process – if for any reason you should abandon (in writing) your patent application prior to the beginning of any of these tasks, the USPTO is supposed to refund the associated fees.

5 Complete the next section down to compute any extra costs associated with claims numbering over twenty, and independent claims over three. You don't get to subtract if you have less than twenty total claims, or less than three independent claims; the minimum value for those totals is zero. Don't miss the line on the right if you have multiple dependent claims. Use Multiple Dependent Claim Fee Calculation Sheet Form PTO/SB/07 to compute the total for your multiple dependent claims.

6 On the next line down, there's an excess page charge for every fifty sheets over one hundred. Very few patents are this long, unless they contain a computer program listing.

7 The last section is for any other fees, which would mostly be late fees (not an issue for first filing), but they give the example of the added fee if your application includes an original specification in a foreign language. In this section, I do recommend you add in the $300 publication fee, as it will usually come due before you hear anything anyway, and is refundable if you abandon your application before that point. On the web form, the publication fee is specifically called out, and there is an additional drop-down menu for petition fees, etc., which most applicants can just ignore.

8 Fill out the signature/date/telephone number section at the bottom (PDF-version only).

9 Lastly, now that you've computed all your fees, don't forget to go back up to the PDF form title box and enter the total on the line provided. On the web form, clicking the "Calculate" button at the bottom will fill in all the math for you.

Action Step 21

(Optional – for paper filing only)

Go to the USPTO's forms website, select PTO/SB/17, and fill it out.

http://www.uspto.gov/web/forms/index.html#patent

Print two copies when finished.

10.2.3 Application Data Sheets

Ordinarily, you won't need an Application Data Sheet.

As discussed when we came to this optional line item on the Utility Patent Application Transmittal Form (PTO/SB/05), keep it simple. You shouldn't need one of these.

10.2.4 Your Completed Documents

Next come the application documents you completed in the Application Workbook.

Next in the assembled application will you place the documents you created in the preceding chapters, starting with the specification, followed by the CLAIMS, ABSTRACT OF THE DISCLOSURE, and Drawings, in that order.

Sometimes you'll see "specification" used in a context including the CLAIMS and ABSTRACT, sometimes not.

So are your CLAIMS part of the specification or not? On the application transmittal form the CLAIMS are not called out separately, but the USPTO directs you to list them separately on the receipt post card (which we will discuss a little later). Don't sweat it, life has greater mysteries to which you could be devoting your energy. Ditto for the ABSTRACT OF THE DISCLOSURE, except the USPTO doesn't actually instruct you to call this out separately on either document. But, since it sits on its own page like the CLAIMS, I treat it as separate on the receipt post card.

10.2.5 Declaration for Utility or Design Patent Application Form PTO/SB/01

The Declaration documents the basic who, what, where for your application.

This is the basic top-level patent application form which declares the applicant(s), type of patent you are seeking, the title of your invention, and a few other details. Most of the fields on the form are pretty self-explanatory. For the typical independent inventor filing his own initial application, most of the form will be left blank.

1 Check the "Declaration Submitted with Initial Filing" on the upper left underneath the form title.

2 On the upper right set of boxes next to the form title, fill in the "First Named Inventor" and "Filing Date" fields. If you have more than one inventor, you'll need also to fill out and append the supplemental inventors section (A) of PTO/SB/2.

3 In the big box in the center of the first page, type the title of your invention as it appears at the top of your specification and check the "is attached hereto" box.

4 I don't recommend checking the "Authorization to Permit Access to Application by Participating Offices" box on the bottom half of the first page. All it does is invite someone somewhere else in the world to attempt to interfere with your patent application by claiming priority, particularly now that the U.S. has joined the majority of the rest of the world as a first-to-file nation (see 2.2.1). If gymnasts in China can suddenly get two years older, you can bet patent applications can do the same thing (I'm not aware that China is currently "participating", but that can change).

5 Page two is exclusively dedicated to referencing either foreign patent applications that you have already filed and want to call out for the purposes of transferring their priority dates to your US application. You also should call out here any applications you are aware of that have already been filed elsewhere for the same invention, but which do not establish priority for you (in this case, check the "Priority Not Claimed" box. If you have anything to fill out here, it may be wise to enlist the help of an attorney. Heaven forbid you need more space for this, but there's a box on the bottom you can check which tells the patent examiner you are continuing this section on a form PTO/SB/2. You can find just a little further down in the directory if you need it (the form calls out PTO/SB/2B – that means Section B of the form; PTO/SB/2A and PTO/SB/2B are just different parts of PTO/SB/2).

6 At the top of page 3 you again (you already did this once in Utility Patent Application Transmittal Form PTO/SB/05) specify to what address you wish the USPTO to send any correspondence related to your application. Just check the "Correspondence address below" box and fill out the fields. After all, if something's worth doing once, it's worth doing twice (ancient Microsoft Office™ users proverb).

7 The final section at the bottom of page 3 is where you list yourself as the inventor. If you have more than one inventor, you'll need to fill out the Section A of form PTO/SB/2B. Ignore the "A petition has been filed for this unsigned inventor" box unless for some reason you lack the ability to sign in the appropriate field. I have never had a need to learn anything about filing a non-signature petition – if you need to do so, the USPTO's website will have the information you need.

> **Action Step 22**
>
> Go to the USPTO's forms website, select PTO/SB/01, and fill it out.
>
> http://www.uspto.gov/web/forms/index.html#patent
>
> Print to PDF. If filing by mail, print two copies when finished. Remember to also complete the relevant parts of PTO/SB/02 if you have more than one inventor, or need more space for callouts to foreign patents.

10.2.6 A Nucleotide and/or Amino Acid Sequence Listing

The USPTO's website and your example patents are your best resource for formatting very field-specific application elements such as this.

OK, I left this in just so there would be one section for each deliverable listed at the beginning of the chapter, but not being a molecular biologist, I don't really have any specific recommendations. Like me, you may find it odd that this one particular type of attachment is called out so specifically by the USPTO (whereas computer program listings, etc. are not) when nucleotide and/or amino acid sequences have nothing to do with 99.99% of all patent applications filed, at least at present. Of course, in the future, who knows? (Maybe someone involved really liked Blade Runner.) For guidance on very field specific application elements such as this, the USPTO's website and the example patents that you pulled aside are your best resource.

10.3 SUBMITTING YOUR APPLICATION

Filing through EFS-Web will save you $400.

If you've completed all of the steps up to this point, you are, at last, done. All that remains is for you to file the application. The USPTO accepts applications both by mail, and alternatively (and preferably) through an electronic filing system known as EFS-Web. Note that with the passage of the Leahy-Smith America Invents Act, submissions via conventional mail must pay an additional $400 surcharge (a word with which congress seems rather enamored of late); thus I now recommend electronic submission. The good news is that submission through EFS-Web differs very little from conventional filing except that you must complete a few very simple web-forms where you upload all files in PDF format. With the exception of Fee Transmittal Form PTO/SB/17, which is incorporated directly into EFS-Web, the standard forms covered in this chapter must all be completed exactly as for conventional submission, the only difference being that you must print them to PDF files instead of paper. (While the USPTO does provide some forms in a format that can be filled out online and electronically saved as editable PDF's, none of the basic filing forms are yet included among them.)

10.3.1 Electronic Filing through EFS-Web

The USPTO's EFS-Web is actually quite slick and contextually populated with a good set of informational links, so the answers to any questions you may have as you work are generally only a click away. Another great thing about this web form which you don't usually see is that there is a row of tabs across the top that allow you to easily skip backward to check/correct previous pages if you suspect/realize that you make a mistake. Here's EFS-Web's URL:

http://www.uspto.gov/patents/process/file/efs/index.jsp

Only EFS-Web registered filers can save a partially completed submission and come back to it later.

Your first step will be to decide whether to file as an EFS-Web registered versus unregistered filer. An unregistered filer must complete the submission in one sitting, which is not at all difficult as long as you are organized. Registering will cause the USPTO to issue you a user number and digital certificate such that you may save a partially completed submission and return to complete it later. Registering also gives you access to the Patent Application Information Retrieval (PAIR) system, which will allow you to track the status of your application online. Registering is a bit of a hassle and not particularly quick, though, involving filling out two forms, one of which you must have notarized and submit by conventional mail. Thus, registering has its pros and cons, which generally I think favor becoming a registered user, provided you apply well ahead of when you plan to submit your application – I wouldn't generally delay filing just to await the completion of the registration process. Information on becoming an EFS-Web registered filer may be found here:

Becoming an EFS-Web registered filer isn't something you can do quickly at the last minute.

http://www.uspto.gov/patents/process/file/efs/guidance/register.jsp#heading-2

There's also a link to this page on the EFS-Web form.

To begin, if you have registered "Launch EFS-Web Registered eFiler" and load your certificate and type your password in the long-in box, or, if you have not registered, just click the "Launch EFS-Web Unregistered eFiler" link. While the EFS-Web is fairly straightforward, the electronic form has a number of pages, so you should budget two hours to complete it (which must be contiguous if you elect to submit as an unregistered user. Here are a few salient tips to expedite:

EFS-Web will send you an e-mail confirmation of your filing.

- The USPTO will send confirmation of receipt of your application to the email address you enter on the first page of the form, so select appropriately.

Expedited filing is generally expensive and contrary to your interests.

- Once you select "New application" and "Utility", a number of filing options will appear. Generally, you should select "Nonprovisional Application under 35 USC 111(a)". I do not recommend the "Track I Prioritized Examination – Nonprovisional Application under 35 USC 111(a)" option, which allows you to expedite the processing of your application, principally because it costs $2,400 extra (double that for large entities), but also because generally it is preferable to use the time between filing and prosecution of the application to pursue whatever marketing strategy you choose (e.g. licensing) before any additional fees come due. The "Accelerated Exam" option is similar, except that it petitions expedited processing for no extra charge, presumably on the basis of one of a number of legally proscribed justifications, one such accepted reason, for instance, being if an applicant has a terminal illness which makes him/her unlikely to survive long enough to see the normal process through to the end (seriously). A relatively small percentage of applications qualify for accelerated examination, which adds paperwork and, again, generally acceleration of the application process is antithetical to your objectives, so unless you have a very specific and rather compelling reason, it's not for you.

- The payment form on the website closely mirrors the traditional PDF form used for filing by mail (PTO/SB/17), and you can reference the step-by-step instructions in Section 10.2.2 to help you fill it out if needed. After the "Calculate Fees" form, the "Confirm and Submit" form will offer you options to either pay now or later, but there's really no benefit to waiting (your application will not be processed until you have paid), and immediate payment is the most assured way to avoid incurring late fees through procrastination. While most will find it easiest to pay with a credit card, EFS-Web also accepts payment via a USPTO Deposit Account (regular users such as patent attorneys have these) or direct electronic funds transfer.

You can pay your filing fees directly through EFS-Web, which obviates the need for Fee Transmittal Form PTO/SB/17.

Application documents can be uploaded in a single file or in any number of files containing any combination of the documents provided you are careful to denote the contents of each file on the upload page. Section 5 of the Application Workbook includes a checklist to help you prepare for and methodically execute the upload procedure. The USPTO accepts a wide range of PDF standards, and you will receive an immediate automatic notification if the format of one of your files is problematic. In a pinch, simply printing hardcopies and scanning them as 300 dpi or better PDF images is acceptable, and actually necessary for form PTO/SB/01, which requires a signature. Fortunately, scanners are generally available in public libraries and retail copy centers if you do not have personal access to one.

EFS-Web accepts a variety of PDF standards, and will warn you immediately if a file you upload is not compatible.

Here's a list of PDF writers the USPTO has verified EFS-Web Compatible:

Adobe Acrobat Professional	**Open Office (Freeware)**
ABXPDF Writer (Freeware)	**PDF Redirect (Freeware)**
Cute PDF (Freeware)	**PDF 995 (Ad ware)**
Easy Office (Ad ware)	**Primo PDF (Freeware)**

I recommend Cute PDF, which is very simple. General information regarding PDF format requirements and settings may be found here:

http://www.uspto.gov/ebc/portal/efs/pdf-creation.pdf

On the EFS-Web resources page there are also links to simulated versions of both the registered and unregistered filer versions of the submission procedures which you can use for practice and to familiarize yourself with the submission forms prior to actually uploading your application:

> **EFS-Web has practice forms so that you can browse/rehearse the submission procedure before going live.**

http://www.uspto.gov/patents/process/file/efs/guidance/index.jsp

While the actual submission process can be cancelled any time prior to hitting the submit button, the practice forms have the advantage of not requiring you to complete each page before moving on to the next by using the "Navigation Simulation" links next to the deactivated actual navigation buttons near the bottom. The practice versions are not currently fully updated with the most recent changes to the live forms, by the way, but they are very similar.

10.3.2 Submission Via Conventional Mail

With the imposition by Congress of the $400 penalty for paper filing, I really encourage you to consider EFS-Web, but the old-fashioned way still works. If that is your preference, print out two copies of the deliverable sections of the Application Workbook (one to send, one for your records), and replace the slipsheet with the forms (which you should have printed when you filled them out). You'll need a full-size envelope (don't fold your application), and a way to figure out the correct postage (it's not the same as a regular letter – a trip down to your local post office will be easiest unless you have reason to be familiar with postage rates).

Address your package to:

Commissioner for Patents
P.O. Box 1450
Alexandria, VA 22313-1450

> **The USPTO recommends you include a self-addressed, stamped document receipt postcard with your application.**
>
> **The best reason to include a document receipt postcard is that you will find out much faster if something in your application goes missing.**

Filing by mail comes with one additional little task. It is recommended that you attach a self-addressed and stamped post card listing the entire contents of your application with your filing. The receiving party at the USPTO will use this card to verify the contents of your application are complete. The postcard will be returned to you, by mail, date-stamped to confirm receipt of all of the listed documents, or indicating what is missing.

There are two reasons that doing this is a good idea. Firstly, the value of a patent is very much about its date of priority, which, with the passage of the Leahy-Smith America Invents Act, is synonymous with the verifiable filing date. Of course, the USPTO will record the date on which they received your application, but people are human and mistakes happen. Speaking of mistakes, what if something gets inadvertently separated from your application at the USPTO? That's the second reason to include the document receipt post card. If you have it as proof that your application was complete upon receipt at the USPTO, you can replace materials in the fairly unlikely event that they should go missing without paying late fees.

On one side your postcard should include:

- The title of your invention exactly as it appears on your specification
- The name of the inventor or inventors

Followed by a space, and then:

- The title and number of pages of each USPTO form included in your package
- The number of pages of your specification (excluding claims)
- The number of claims and the number of claim pages
- A callout for your abstract of the disclosure and page number (which should be one) (since it's separated from the specification by the claims in the suggested ordering on the USPTO's website)
- The numbers of figures and sheets of your drawings
- A callout that a declaration statement is included and the number of pages
- The type and number of any other documents that are included and the number of pages of each document.
- The amount of payment and the method of payment (e.g.: check, credit card, money order, or deposit account)

List contents by their order in your application package.

Naturally, it's best to list them in the order they appear in the package. This will look something like the example shown in Fig. 10-1 below:

METHOD OF MAPPING LOCATION OF STOLEN BOOTY
Inventors: Captain Joseph Flint
            Billy Barnacle Bones

___ Utility Patent Application Transmittal Form PTO/SB/05 (1 page)
___ Fee Transmittal Form PTO/SB/17 (1 page)
___ Check for $545.00
___ Specification (15 pages)
___ 19 claims (3 pages)
___ Abstract of the Disclosure (1 page)
___ Drawings totaling 6 figures (4 pages)
___ Declaration for Utility or Design Patent Application Form PTO/SB/01 with PTO/SB/02A supplemental inventor sheet (4 pages)

Fig. 10-1 Example Document Receipt Postcard

The opposite side should include:

- Your name
- The address to which you want the receipt post card sent
- Correct postage

The easiest way to go to a postcard format is to just print a label and glue it to a postcard.

You can use the template provided in the workbook, but you will have to figure out how to get it printed on a postcard. I like just printing it on regular printer paper, cutting it out as a label, and gluing it to a card. Alternatively, there's nothing wrong with just writing it on a postcard by hand, if your penmanship is still legible to others (print, don't use cursive).

The USPTO directs you to "attach" the postcard to the first page in your package, which will be the Utility Patent Application Transmittal Form. I interpret that to mean a paperclip.

Action Step 23 (Optional – for paper filing only)
Make a document receipt postcard for your application. A template is provided in Section 2 of the Application Workbook.

10.3.3 Congratulations!

Now that you're a veteran, your next patent will be considerably easier.

Take a moment to think about what you have accomplished – it is no small thing...but, I'll bet that before long you'll realize that this first application is just the beginning, and even as soon as your next endeavor you will find the patent application process much easier. As a matter of fact, many users will already have identified and be chomping at the bit to start the next project – that's just the nature of inventors. For a true inventor, the best thing about finishing one project is that you now have time to start the next one.

10.4 OTHER RESOURCES

Part of everything you undertake should be to continuously better your skills and knowledge.

You should never stop learning – a part of anything you do should be continuous improvement, and patent writing should be no different. Having completed a patent application using this guide, you should now possess a basic understanding of the advantages, purpose, mechanics, and strategies of patent writing. Obviously there's more to learn, and additionally, it is to be expected that an occasional user of this guide may have an unusual need a little farther off the beaten path. To these ends, the following is my short list of recommended other resources (note that, as everywhere in this guide, the links are presented as URL's so that hardcopy users can type them:

The United States Patent and Trademark Office (USPTO)

First and foremost, the USPTO provides extensive resources for independent inventors:

http://uspto.gov

To go to the USPTO's page containing "General Information Concerning Patents":

http://www.uspto.gov/web/offices/pac/doc/general/index.html

To go straight to the USPTO's "Guide to Filing A Non-Provisional (Utility) Patent Application":

http://www.uspto.gov/web/offices/pac/utility/utility.htm

The USPTO's guide is not nearly as step-by-step as this one, but all the basics are there.

Google

Another resource you should never overlook is Google. The Google search engine sees everything that wishes to be seen, and some things that don't. In Section 10.2 I suggested Googling those miscellaneous government document references that you come across, but you can do the same thing for any keyword or phrase that you don't immediately understand. As often as not, a Google search on patent-related keywords will take you straight to the page you want to see on the USPTO's website. Other times, you'll discover an alternate source.

Brown & Michaels Intellectual Property Home Page

Brown & Michaels provide a good set of free and comprehensive resources that I have found occasionally handy. Here's a link to their home page:

http://www.bpmlegal.com/index.html

You may find particularly useful their page of "Pitfalls and Traps in Claim Drafting":

http://www.bpmlegal.com/howtopat7.html

Finally, I recommend you take a look at their "Weird and Wonderful Patents" page, just for fun. Hopefully, if you've applied the guiding principals of Chapter 2, you won't soon be adding a patent to this list:

http://www.bpmlegal.com/weird.html

Patent It Yourself by David Pressman

The Essential Inventor's Guide provides very practical nuts-and-bolts information from one inventor to another precisely on writing a patent application in MS Word (or equivalent). Although you may appreciate it far less, a patent attorney's perspective will also be invaluable to you. *Patent It Yourself* by David Pressman is an excellent and very reasonably priced resource that will serve you well in this regard. Now, if you think this guide was a bit much to read, be forewarned that the 14th addition of Pressman's book is 596 pages in length, but additional information it includes will assist you in the further prosecution of your application, such as answering office actions (advice for which the best source being a lawyer), should you need to do this on your own. (I caution you, however, to reevaluate the potential value of your invention if you have not succeeded in licensing it by the time you need to respond to the first office action.)

11 Licensing Your Invention

11.1 Selecting a Licensee

Once you file for your patent application, there's no time to lose if your goal is to license your invention.

Most users of this guide will intend to license their invention to a third party. To maximize the time you have to succeed, the process of finding a licensee should most certainly commence immediately following the filing your patent application; thus, preparations for this key step may be concurrent with the development of your patent application, and there naturally are some synergies that may be realized. Nevertheless, licensing your patent is largely a separate and independent endeavor, comprising its own set of strategies, facilitations, and pitfalls.

Do not feel intimidated. Having completed the steps in this guide, you're well qualified to be in the market.

Now, looking out into the big world around you, you may be tempted to feel small and outgunned, and only hope that you can get someone, anyone, to take you seriously. Assuming you have objectively performed the evaluation steps of Chapter 2, such meager self appraisal could not be less justified. You are not asking anyone for favors. You are the real deal. You have something genuine to offer, and you aren't trying to fool anyone. Conversely, you aren't proposing to grace a potential licensee with a special privilege beyond words, either. You are a business person looking to get together with other business people and make money. Nothing more, nothing less. Be a straight shooter, tell it like it is, and you will be respected by anyone whose respect is worth having.

SELECTING A LICENSEE

Be very judicious in your selection of a licensee, your project will not likely survive a bad choice.

Let us begin by discussing whom you should be considering as a potential licensee. You are not desperate, and you should be downright picky about who you propose to work with. An ideal licensee needs to have both the ability and motive to meet your objectives, and all of your thinking should be framed in terms of these two attributes. Below is a list of typical features by which you may compare/evaluate your prospects:

11.1.1 Size

Large organizations typically have more resources but less motivation to advance a new product.

With respect to potential licensees, size does matter. The size of a company can affect both ability and motivation to make something happen. Clearly the development resources and market reach of a large company have much appeal, but actually bringing them to bear upon your project can be difficult, even if they license your invention. Larger companies tend to have a greater number of parallel development projects simultaneously ongoing, and so your project will typically be one of many. They don't need any one project to succeed (indeed, they may not need any). Ideally, it would be in your best interest for whatever licensee you engage to put all of their eggs in one basket (yours), but, of course, their interests only partially overlap yours.

Product development often serves a more defensive than offensive role for leading market shareholders.

So an industry leader certainly can bring your product to market, and with their marketing resources can virtually guarantee a certain level of success; the only question is, will they?. Very large companies are inherently more bureaucratic, conservative, and generally less hungry for market share. They're happy the status quo, and not looking for a change. Their product development activities stem more from defense than offense, as they already own plenty of the market – they just don't want to get left behind. Frankly, from a defensive perspective, obtaining an exclusive license on your invention and sitting on it is just as good as developing it. Now, within such organizations, there will always be young bucks looking to get ahead, and they're plenty motivated. But the real decisions are made above them.

When choosing between ingredients for success, I'd bet on hunger over fat resources every time.

Clearly then, you're looking for a balance; however, favor a smaller, rather than a larger licensee. At the end of the day, drive and a genuine need-to-succeed can make up for any lack of resources, but the converse is certainly not true. If your product really is good, a motivated licensee has all the resources it needs.

11.1.2 Competing Market Share

Regardless of size, pay attention to any potential conflicts of interest with a company's existing products.

Clearly there certainly tends to be some overlap between this category and the last, since large companies tend to be so because they have good market share; but do not take the two as exactly synonymous. A company may be large, but that does not necessarily mean they have large market presence in your particular product arena. Conversely, a small company may make only a single product for which yours is a direct competitor.

> **Do not license to an organization that is highly invested in a product with which your invention would compete.**

If necessity is the mother of invention, for a business innovation is born of competition.

So doing would be begging them to just sit on your invention for a few years doing very little while they first pursue their established first string for all its worth.

Note also that a competitive marketplace is your friend while you attempt to license and have your invention developed into a fielded product. The more competitive the environment, the more motivated a manufacturer will be to launch new (and even multiple) products. In an ideal situation, you want a well-established and successful licensee who needs to increase competitiveness in a single product area that coincides with your invention. You want an organization that is hungry to find an edge and think your product has a good chance of being just that. Of course, once your invention becomes a product on store shelves, you'd love it if all that pesky competition just went away, but unfortunately you just can't have your cake and eat it too.

11.1.3 Heritage

It's hard to beat a prolific history of licensed product development.

A potential licensee may be able to make a good case for themselves if they can show more-than-adequate technical know-how, facilities, and capital to develop your invention. Nevertheless, the best evidence of all is simply a history of successful, similar endeavors, both in terms of successful internal research and development, and, especially, other licensed products they have lucratively developed and marketed. Of course, not every candidate that can't show such a history is a dud, since companies really do sometimes change business strategy and focus, but clearly if they don't have a track record, you should have a tangible and credible explanation for why that would change.

More than anyone else, recent start-ups are likely to give your project their all, but they may also give up the ghost.

Obviously, one of the least impeachable excuses would be if they are a relatively new company. For a fairly recent start-up motivation is generally high; but, as newer enterprises are largely untested, the risk associated with their ability to follow through is equally high. Therefore, in such cases you should focus your evaluation on the company's viability, which will be effectively a composite of its capitalization and the capabilities of the management and technical staff.

> **Beware the "not invented here factor".**

Perhaps the best reason to favor potential licensees who have shown a history of developing products from licensed patents is that it provides evidence that an organization has what it takes to overcome what I call the "not-invented-here factor". When you approach a potential licensee, you will often see two faces. The first will be that of a business development professional, and this person may be genuinely very interested in finding out about what you've been doing and whether or not you can make money together. That's part of his/her job. The second face will be someone within the authority structure of their product development arm. This person often won't seem to be on the same page as the business development lead, because he's not. His (or her) team's job is to develop new products, and their true focus is to invent them internally. He has no interest in having an outside inventor show up with what he regards as something his team should have invented first, and he'll be quite skeptical that your idea could be better than anything they're already looking at…and by "skeptical" I mean not interested, or, more specifically, interested in having you dismissed.

That business development is interested in your idea doesn't guarantee a prospect's technical staff wants to play ball.

It follows, then, that while certain inefficiencies exist for organizations that subcontract out development work, there exists one distinct advantage in that they won't feel the least bit trumped if your invention really is the next big thing. For the hired guns actually performing the development work, this will be business as usual. I'm not suggesting that you should favor such prospects, but you should recognize that their reduced internal resources do come with one great strength. Inevitably, product developers are usually also inventors (at least at heart), and there always exists the potential for someone in the mix to instinctively feel the need to compete with your invention rather than embrace it, regardless of what they're being paid to do.

Inventors at heart, product developers will instinctively want to beat any invention they see.

Realize that in the world in which you live, in spite of what anyone trying to get elected says, most people pretty much are team players exactly as long as the team is serving their best personal interests. You may not want to accept that; cannot accept that; won't accept that – you can believe whatever you want; but, let me at least try to convince you to pretend what I say is true at least long enough to license your invention.

Most people place their own interests ahead of their employer's.

Fortunately, not everyone is like that (many or even most, but not all), and you're looking to work with the good guys. Be very alert to the not-invented-here factor. If you sense that a prospective licensee's technical staff doesn't want you there, you should really go somewhere else. As I said above, you won't initially see the technical face, and business development will sincerely act in their organization's best interest to cultivate a relationship with you. Business development professionals are often blissfully ignorant of the not-invented-here factor, and may inadvertently simply waste your (and their) time in a process that may ultimately prove pointless. When their technical staff can't seem to get around to evaluating your proposal, they'll earnestly do their best to keep you on the line, not wanting the opportunity to pass them by without objective evaluation.

Business development professionals are often oblivious to the not-invented-here factor.

In your initial negotiations, one goal you should set is to arrange to dialogue with the technical personnel (internal or external) who will be responsible for productizing your invention as early as possible in the process,

Your first order of business in negotiations should be to get a candid assessment from the technical staff that will be responsible for developing your invention into a product.

ostensibly to receive a candid evaluation of your invention's prospects and what technical hurdles may be involved, but, in fact, equally to serve as an interview of those that would be in control of your invention's destiny. If they're not fans of your invention or simply of working with you, that will quickly become apparent. If you cannot get their technical representatives/associates to make contact with you for whatever reason, that's a bad sign. What you want to avoid is several wasted months of dialogue followed by an apologetic dismissal based on a final assessment by someone to whom you've never spoken, but you realize has simply been stonewalling the business development lead until he runs out of breath; or, even worse, flying out for a meeting where you are cordially met by enthusiastic business development representatives, but where the technical staff stand you up (both of which I have the distinct pleasure of claiming as personal experiences).

Failure to get buy-in from an organization's key technical personnel usually spells doom for a project.

Do not fail to recognize that two critical steps remain following licensing of your invention, those being successful development of a producible product, and then effective marketing. Even if you manage to make such a compelling case for your invention that a company's decision makers overrule internal objections/lack of interest, realize that failure to get buy-in from key technical personnel is generally fatal to a project, regardless.

A dozen bad prospects + one good one equals success if you can figure out which is which.

Let me assure you that mediocrity poses a far greater obstacle than incompetence. Now, that is indeed a whole lot of negativity, and I do not want you to respond with undue apprehension. The goal is to make you savvy, not gun-shy. Remember, after all, you're the one who's aligned with that company's best interests, not the stone-waller who needs to grow thicker skin, but you can only choose how thick your own skin is, of course. At the end of the day, if you locate ten companies that turn you down, another five that would make poor licensees, and one excellent choice, only the latter matters at all, as long as you have the aptitude to figure out which one it is. Inventors forge successful alliances every day.

11.1.4 Knowledge Relevant to Your Field of Invention

Select a licensee with relevant know-how.

It isn't realistic to expect a licensee to thrive in a product area where it has no existing expertise.

Clearly a licensee familiar with similar devices will productize your invention with greater alacrity than one who hasn't a clue. This does not mean the company must already be involved directly in your product area. If your device is a special type of metal grinder, you want a licensee that has experience making power tools; if you've come up with a new kind of breakfast cereal, you want a licensee with experience marketing food products, etc. Within any industry, there exists a multitude of little, but rather important, details about which only experienced veterans will know anything. Do you know what hoops you need to jump through to get approval to market a food product in the U.S.? I don't, but I bet there are quite a few. Don't expect anyone to succeed if they have to reinvent the wheel.

Keep in mind, none-the-less, that a company does not have to internally possess all of the required resources, but they must have access to them. As I mentioned above, sometimes it works out better if some of those resources are contracted. Realize also that even strictly technical *savior faire* goes well beyond the nuts-and-bolts of understanding how to optimize your invention as an individual unit. To succeed, your invention must become a mass-produced article, and knowledge of manufacture and well-greased manufacturing relationships are often more critical than specific technical expertise, which can be more easily contracted.

Knowledge of manufacturing processes and supplier relationships is just as important as product development capability.

11.2 Finding Prospective Licensees

So how do you find prospective licensees? In the modern world I really have only one answer. You guessed it – the internet. Anyone who's serious about doing business (at least the kind you're interested in) has an internet presence. If a company doesn't, have nothing to do with them. Would you really want to go with a licensee living in the stone age? (But, I'm sure you are aware that the converse is not true – a flashy website in no way portends a company's legitimacy, only the lack thereof.) Generally, most inventors will be familiar with the major players in the relevant industry, and so you probably already have a few options in mind; but, always remember the market leaders aren't necessarily your best candidates.

The best source of contact information on prospective licensees is the internet.

Basic keyword searches on a search engine usually dredge up plenty of stuff, but most of what you will see are retailers, and you seek manufacturers. So, for the first time in your life you may find yourself scrolling through more than the first two pages of search results listings. Retail sites are still useful, nevertheless, because you can browse product listings for brand names, and a subsequent search on a specific brand name will usually bring up the manufacturer on the first results page. Note that brand names are often not synonymous with the manufacturer's name, however, so the manufacturer won't always come up. Moreover, even if a manufacturer site does come up bearing that brand name, it still may not reveal the actual manufacturer's name, as many host separate websites for each of several brands. Of course, all of a manufacturer's websites will have contact information, but you are looking to place your solicitation as precisely as possible into the mailbox of someone who will be in a position to do something with it. Also, you'll want to address your solicitation to the manufacturer's actual name, not one of its brands, which will make you look like you haven't done your homework.

Manufacturer's names are often not synonymous with their brands.

Many industries have trade organizations whose websites provide a portal to their member manufacturers. If you come across one of these, bookmark it immediately, as you have found a very efficient means of getting to exactly the data you want..

Trade organization websites often provide links to member manufacturers.

Be selective.

Don't just contact the first manufacturers you see, and don't contact more than you can juggle all at once.

You should be selective. You'll want to initially contact about five to ten good candidates, but don't pick the first ones you come across. Go to their websites and look at their product offerings. What you want is to find information that speaks to relevant capability and motivation. In particular, what you want to see are products in family with your invention (but without too much overlap). For instance, let's say, hypothetically, you've invented a new kind of fast response valve. There are many types of valves on the market, designed for different types of fluids, operating pressure ranges, flow rates, response times, service life, etc. If you started surfing the web, you'd quickly realize that manufacturers specialize, and one that makes ultra-clean, lubricant-free valves for pharmaceutical dispensing will have much less interest in your valve than one who specializes in high-speed valves. As you search, make a list, ranking them in terms of how their market focus fits with your invention.

Update your list of prospects as you go.

As you learn from experience, your priorities may change.

Once you've started soliciting potential licensees, depending on how much interest you generate, you may need to supplement your initial contacts with others further down your list, which shouldn't be static. As you dialogue with your initial contacts, you may learn a few things which cause you to reprioritize your candidates, certainly the ones to whom you've already spoken, but also those which you simply ranked based on earlier assumptions that may have changed.

11.3 MAKING CONTACT

Cold calls don't work much better in the business world than they do at home.

So how do you make contact, once you have your list of candidates? Well, there are no set rules, but realize exactly what you are doing – one way or another, you're making a completely unsolicited proposal. Recall for a moment how you respond to unsolicited proposals. When you receive a call at home or work from someone trying to sell you something (even worse, these days there seems to be an ever increasing supply of fantastic ideas of how you can just give your money away), what do you do? Smart people hang up immediately. The too polite focus on nothing more than excusing themselves from the call while the telemarketer (jerk) reads questions designed to take advantage of their good nature to keep them on the line. Even if you hear something interesting, you are very apprehensive if you have any sense about you at all. And then there's the fact that getting directly through by phone to anyone who matters may be even more challenging than getting them to listen to you once you have. So, I don't recommend cold calls.

A good solicitation is welcome relief from the dull and ordinary.

Initial contact by written media has distinct advantages, the most significant being that you can very carefully tailor the pitch to keep them reading, and if you do it right, it can certainly be much more interesting than most of whatever else lurks in their almost assuredly dull heap of mail (conventional or electronic). Whereas a phone call is almost universally an unwanted interruption, a well-presented solicitation can be a welcome diversion. Also, the phone call leaves them with nothing tangible, but they can immediately show your written solicitation to colleagues. Frankly, between email and conventional (snail) mail, I quite prefer the former, but both have advantages and disadvantages, depending on the situation.

Snail mail may be preferred if you cannot obtain email for the appropriate contact, but you know the person's name. To get snail mail to the intended recipient, all you need is the company's mailing address and a name. Obviously, email requires much more specific knowledge. Also, in a pinch, snail mail may fare better if addressed to "Business Development", vs. email to a generic customer contact. Between sending an email to a less relevant recipient, and landing a direct hit with snail mail, choose the latter. Don't make them resort to snail mail to respond, though – always direct correspondence to an email address.

Conventional mail may be your best bet if you wish to target a specific recipient but can't obtain an email address.

Email is the modern way of business communication, but electronic mailboxes tend to be over-spammed. You can usually avoid an automated spam filter if your email is addressed directly to a single recipient (which it should be), or sent to a contact email address specified on a website; but, even still, most of us are drowning in a swollen sea of incoming traffic; so an email may get lost in the shuffle. Email does have the distinct advantage that it makes responding to you effortless, and if you have missed the mark, for the recipient to forward your message to the correct contact person (and wash their hands of it), is equally easy.

The advantage of email is that it makes forwarding your solicitation to others within the organization effortless.

Remember that the personnel of interest are busy, and the difference between clicking forward/send and having to stop what they are doing and get out their chair can be the difference between your solicitation finding its way to real consideration or the circular file. If at all possible, you should try to direct your email straight to the correct person within the organization. For smaller companies, that'll be the CEO. For larger organizations, look for a business development contact.

Your email should be as targeted as possible.

If you can't find out on your own, there's always the two-step process of first asking where such a proposal should go from a more generic contact that you can find. If you're just trying to locate the correct contact to whom to address your proposal, the phone is OK – just don't try to market your invention starting with a phone call. Be ready, though. Depending on the circumstances, you might be transferred straight through. In that event, be prepared to explain briefly who you are and that you have some information that you believe would be of interest to their organization. Your goal remains to get your solicitation to them.

There's no harm in using the phone to find out to whom you should send your solicitation.

You're approach should be surgical, not shotgun. Do not spam more than one contact within a single company with your solicitation. That will make you look sloppy and desperate (kind of like all those other e-spammers). If your invention is good and well presented, you should expect to receive responses from many, but not all of the addressees. If, after you have sent your solicitation, you hear nothing by the expiration of your specified time-to-

Spamming more than one contact within a single company with your solicitation will make you look sloppy and desperate.

respond (we'll talk about that later) from one or more recipients, it means one of three things:

- Your proposal has been rejected and the company hasn't the courtesy or internal machinery (i.e. there may be no one who considers it their job, so they're all assuming someone else will do it) to provide you a response.

A no-reply never reflects well on a company.

- Your proposal didn't get evaluated because it never made it to the right person.

- They're still trying to get around to it (it's sitting on someone's desk).

Any company that would make a good licensee will respond to a good opportunity.

Let it not come as a surprise that some very successful companies would make poor licensees.

Obviously, if you've heard nothing, you won't know which is the case, but regardless, none of these reasons are a good sign. (The second bullet is perhaps the least disparaging; but, assuming you didn't send your solicitation to the janitor, still…) Many of those shady experts-for-hire who market themselves as specialists at obtaining licensing agreements will go on-and-on about how you need someone like them to get a company to pay attention to you. The fact is that any organization that you would want as a licensee won't ignore a well presented good idea. If your invention has real potential and a company doesn't respond, there's something wrong with them, not you.

With regard to a particular potential licensee that has received your solicitation in total silence, you now have nothing to lose, which changes the rules a little bit.

In response to a company's no-response you can:

Drop them.	Certainly, if you have adequate responses from other and there's no particular reason to prefer a non-responsive candidate over others, just dropping them from your list is very reasonable. Remember, the first goal of your licensing effort is to weed out bad prospects. A bad licensee is worse than no licensee.

or…

Send your solicitation to a second contact within the company.	Don't send your solicitation a second time to the same contact. Sending your solicitation to an alternate contact within the organization may get a response if your original simply didn't land on fertile soil. If the original submission had been reviewed and dismissed, your second round may annoy the recipient if they were in the loop the first time, but that won't cost you anything – you can't lose what you never had (and assuming they never received the original is fair game if they can't take a moment to acknowledge that they have – you're extending them the courtesy of giving them the benefit of the doubt). Of course, you can only do this if you have identified a second contact.

or...

Do a follow-up call.

If you are going to make a follow-up attempt with a non-responsive solicitee, the best approach is to now pick up the phone. Start with the most direct phone number you have, which, admittedly, often will be the general contact number. Just tell the representative who answers that you're following up on a proposal you sent and ask to be connected to the appropriate member of their staff. They will probably have no idea your solicitation ever came in, nor any knowledge of where it went (if anywhere), but they will usually connect you to the person to whom it should have gone.

That person will either dismiss you or ask you to resend your solicitation directly to them (exactly what you wanted in the first place!). Let him/her do the driving – don't try to talk about your invention on the phone, unless they ask. You want them to read your solicitation, don't you? They almost certainly aren't going to invite you to come talk to them about it in person until they have. Talk is cheap, and it will be next to impossible for you to get anyone really interested in your invention if they haven't seen a picture.

All of these actions have their place, but again, if you objectively evaluated the quality and potential value of your invention per Chapter 2, you'll get responses from the type of organizations that you seek.

Retain all correspondence.

Be organized. You should retain all correspondence to and from each candidate, and they should be compartmentalized, one folder for each contact (be that electronic or cardboard), to ensure that you never mix them up or accidentally send information intended for one to another.

11.4 THE SOLICITATION

Your solicitation should be short and to the point.

Craft your solicitation with the objective to get their attention and convince them that your invention is interesting and real enough to be worth following up on. Do not waste language – keep it short and to the point. Do not say something in two sentences that could be said in one. Everything they initially believe about you will be formed by how you present yourself in your solicitation, and you never get a second chance to make a first impression.

See to it that you present yourself as:

Professional — Professionalism is one of those difficult-to-describe-but-you-know-it-when-you-see-it things. Do not confuse it with being slick or having a fancy presentation. The single best mark of true professionalism is concise, effective communication. Words are not wasted. Neither are graphics, which are not used to decorate, but only where they facilitate communication.

Thorough — Being concise and conveying that your evaluation and development effort has been thorough may seem contrary to one another, but they are not. You do not have to include all of the details to communicate that you've done your homework. Basically, stating that your invention is patent pending, and that you have a working prototype should get that message across.

Grounded — Be a realist. This refers not so much to what should go in your solicitation as to what you should leave out. The truth is a good invention's value will be fairly apparent to those in the industry. Hype and talking up how amazing the invention is will generally be a real turn-off, as will be a lot of discussion of how much revenue it will generate. Understand that the success of your invention may seem like a done deal to you, but from where the recipients are sitting, your solicitation is an invitation to a lot of work and investment with uncertain return. Generally, avoid any talk of expectations in your solicitation, and you won't have to worry about turning people off by expecting too much – if you must, you always have a second chance to come across as unreasonable during negotiations.

Any talk of compensation will be very out of place in your solicitation.

Your solicitation should definitely not make any statement regarding desired compensation or preferred terms and conditions – you don't talk about such things until you have at least given the prospect a chance to express interest.

Your solicitation will have two components – (1) a cover letter, which mostly serves to get their attention and convince them to read (2) a technical disclosure. We will discuss each of these in turn.

11.4.1 Cover Letter

It is vital that your cover letter get the recipient's attention, and convince him/her to read your technical disclosure. It should be brief, and effective even when only skimmed.

Here's a good general cover letter outline:

1 Start with a one or two sentence introduction about why you are contacting them. You don't even have to say anything about who you are – at this point, who cares?

2 Lead into a bulleted list of the benefits your invention provides – not more than ten bullets. Absolutely do not try to describe the benefits in a paragraph format. The reader just wants as quickly as possible to ascertain whether or not they should be interested, and that means simply to know what you have to offer. No details, just the potential benefits. Each bullet should not be more than two lines long, and preferably one.

3 State that you have a working prototype (assuming your invention is of a nature where this is practical and the assertion truthful).

4 Talk about why you selected them as a potential licensee, i.e. what about their organization makes them a good fit for your product. Unlike the other elements of the cover letter, which will be generic, this one to three sentence segment should be specific to the recipient. Your intent here is to impress upon them that you hand-selected them and that they aren't one of five-hundred spamees. Address the recipient and organization as specifically as possible.

5 Lay out the timeline to respond and provide contact information. This will take the form of a polite request (but it will be understood that you are moving on after that date).

6 I recommend a statement limiting the usage of the solicitation materials exclusively to evaluation purposes, and for a finite period of time.

Organize information in your cover letter to keep the reader's interest.

The order here is not arbitrary. The information is laid out such that a typical recipient will encounter each piece of the information according to the likely natural flow of his/her interest; that is, we're effectively answering questions in the order he/she would most likely think to ask them so as to not lose his/her attention.

Here's an example which you may find helpful:

> Mr. Wonka:
>
> I am pleased to bring to your attention the "Everlasting Gobstopper" a revolutionary new technology privately developed over the last two years which has now reached a mature, demonstrable state. This patent pending new simulated candy lozenge:
>
> - Maintains full flavor virtually indefinitely, even with continuous use.
> - Does not dissolve or ablate, permanently maintaining its original size, shape, and color.
> - Yields zero calories.
> - Does not promote tooth decay, but rather prevents it by stimulating salivary production.
> - Can be produced in virtually any flavor.
> - Can optionally be designed to periodically change flavor.
> - Is completely safe, and, in fact, much safer than consumable food products bearing artificial flavors and/or colors.
>
> A prototype manufacturing device has been constructed and successfully tested.
>
> Assessment of the potential benefits of acquiring an industry partner under a licensing agreement for final development, manufacture, and marketing has distinguished the Willy Wonka Candy Company as a current industry leader in terms of market share, product quality, and customer satisfaction. It is, however, the Willy Wonka Candy Company's notable children-come-first reputation which leaves me most convinced that mutual benefits could be realized through a partnership with Willy Wonka.
>
> Please take a moment to review the attached technical disclosure, which describes in more detail the technology from which the Everlasting Gobstopper derives its remarkable features.
>
> Would you or a representative be so kind as to reply to this email by 12/9/11 with any comments/questions and for further information on how the Wonka Candy Company (and its customers) can benefit from this remarkable advancement?
>
> This message constitutes written authorization for Willy Wonka Candy Company representatives to duplicate and use the information provided herein and here attached for evaluation purposes only for up to sixty days from the date of receipt of this message. Thereafter, all reproduced materials must be destroyed unless other authorization has been provided in writing.
>
> At your service,
>
> Dexter R. Sweet

So what did I leave out? Everything technical – I didn't say anything about the fact that the Everlasting Gobstopper is a poly-crystalline plastic ball with a custom nano-sculpted surface designed to stimulate taste buds with no actual mass exchange. Proud as you are of your technical achievement, all of that detail has no place in the cover letter. You're not trying to impress them with your cleverness, but only that your product can provide an advantage in the marketplace.

> **Save the details of how your invention works for your technical disclosure.**

I also didn't talk about additional technical development needed, say, if there are still some glitches to work out. It'll pretty much assumed additional development is required (there's still quite a bit of engineering ahead to go from a prototype to a mass-produced product, even if the prototype functions perfectly), and it works against you to turn them off with the open issues. You're not trying to sweep any product immaturity under the rug, but such details are completely out of place in the disclosure letter.

> **Don't bring up open technical issues; these are expected and unneeded details in the solicitation. Are you trying to persuade or dissuade?**

Remember, in your solicitation you're not asking them to commit to anything. Your entire objective is simply to get them to contact you so that you can open a dialogue with both their business development and chief technical staff, with the ultimate goal of getting invited to present your invention to them in person (if that dialogue convinces you they mean business). Only when presenting in person, after you have had an opportunity to impress them with your prototype, does it make any sense to proceed into conversations about the next development steps. Until they see your prototype, they can't get excited about what you have, and, if they aren't yet excited about what you have, they can't embrace the work ahead with enthusiasm.

> **You'll have plenty of opportunity to be frank about technical challenges once you've given them an opportunity to be interested enough to consider taking them on.**

Notice the example mentions the invention is patent pending. That's important. Don't give anyone an excuse to claim they didn't know and it's your fault. At this point, every document you create should make sure to mention the patent pending status of your invention.

> **Always be sure to note the patent pending status of your invention.**

11.4.2 The Technical Disclosure

While a good cover letter gets their attention, a good technical disclosure document is vital to establishing credibility. Whereas your cover letter contained language specific to the recipient, your technical disclosure needn't. Typically, the most effective level of detail lies somewhere between the BRIEF SUMMARY OF THE INVENTION and DETAILED DESCRIPTION OF THE INVENTION you included in your patent application.

> **A good technical disclosure is your path to credibility.**

Do not send them a copy of your patent application!

Naturally you're proud of your patent specification, but do you really expect a busy (probably skeptical) person to take the time trudge through all that?

Would you?

Usually sending them a copy of the patent application won't do any harm, but it won't really do any good either. In unusual circumstances, your application could face a challenge if the recipient feels there's some overlap with something they already have in the works, or is just plain underhanded. (Of course, in fields embroiled in high-stakes competition/contention over IP, that may actually be more likely than others.) So, there's an off chance you'll do some damage by mailing copies of your patent application, but no real good can come of it – so we're not going to do that. Besides, recall how much fun those were to read when you did your patent search. Using your patent application as a marketing document is like wearing a three-piece suit to the beach. The text is strategically defensive, the figures are separated from the text, every possible technical permutation but the kitchen sink is in there, and there's a ton of extra stuff…and the required formatting hails from the 1800's. It just won't do.

Your technical disclosure should be designed to allow the reader to understand your invention quickly and easily.

Make a technical disclosure document that focuses only on the important features and guides the reader through one to three key figures plus a photo of your prototype (if you have one). Your objective here is to make it easy for the reader to understand the principal technical details of your invention quickly, and the information should be layered so that a basic understanding can be achieved by skimming prior to digesting the entire disclosure line-by-line. This guide is a good example of a proposal-style layered information layout. Maximize use of bulleted lists and graphics to set apart key features and messages. Provide one-sentence "take-aways" along the left-hand side of the text. A reviewer should be able to glean the fundamentals by reading only these paragraph summaries and looking at pictures.

Getting Started – The Cover Sheet

A proprietary information statement should be included on the front cover of your technical disclosure.

Your technical disclosure should have a cover sheet with the name of your invention, and your name(s). I do not recommend artwork on the cover, as it just bulks up the file size and generally looks fluffy. I do recommend a proprietary information statement specifically stating:

- The disclosed material is protected by pending U.S. patent.

- The technical disclosure is provided for evaluation purposes only.

- Separation of any part of the proprietary information statement from the disclosure is prohibited.

Something like the following should do nicely:

> **Proprietary Information**
>
> The contents of this document are protected under pending U.S. patent and provided to [PROSPECTIVE LICENCEE'S NAME] for evaluation purposes only. Disclosure of any of said contents not already known to [PROSPECTIVE LICENCEE'S NAME] prior to viewing to any party except in the specific context of aiding [PROSPECTIVE LICENCEE'S NAME] in evaluation of said contents is prohibited. Separation of this message from or any form of modification of any part of this document without written permission by [YOUR NAME(S)] is prohibited.
>
> The contents of this document are to be considered in all respects exemplary and/or illustrative and not complete or restrictive, the scope of the disclosed intellectual property being indicated by the claims of pertinent pending U.S. patent(s).

A proprietary statement attached to an unsolicited disclosure is pretty toothless, but it's better than nothing.

As with all the examples in this text, feel free to use the example verbatim if you like, or write your own if you have reason to prefer a different style (but regardless, the content should be pretty much the same). While you haven't much legal horsepower (since they haven't signed anything and your submission is entirely unsolicited), you have identified the information as proprietary, which, from a legal standpoint, means you have given fair notice that you regard its outside disclosure as damaging, and that's better than nothing.

The Introduction

The BACKGROUND OF THE INVENTION from your patent application may often be pared down to form the basis of a good introduction.

While you won't be able to use your patent application text wholesale, you will probably find parts of it useful as a starting point. You needn't start in with cold-contact introductory language like "Please allow me to…", or "I am pleased to…", etc. – you're cover letter already broke the ice. Consider beginning with the text from the BACKGROUND OF THE INVENTION you included in your patent application as an introduction, because it makes a case for why your invention is desirable or needed. It will probably need editing to tailor it to your new goal (especially if it's simply too long), but this one part of the patent text should be fairly transferable. You don't have to, however, if you expect your audience to already be acutely aware of the "what would be really greats" in their industry, which really depends upon the breadth and diversity of their product base. Depending on your field/perceived audience, you can opt to just start straight in with something to the effect of "The Everlasting Gobstopper is a simulated candy comprised of a non-toxic crystalline plastic ball with a custom nano-sculpted surface designed to stimulate taste buds with no actual mass exchange." On the other hand, as long as you keep it short, a background-based introduction won't really do any harm. Even readers that find the background annoyingly obvious will sympathize that every document must begin with some form of introduction.

Your introduction should be limited to one paragraph, which should lead directly into the same bulleted list of advantages from the cover letter (which, of course, originated from content in your BACKGROUND OF THE INVENTION, so leave that text out in favor of the bullets). Do not count on everyone reading your technical disclosure to also have seen your cover letter; that will often not be the case.

Definitely include the same bulleted list of advantages you made for your cover letter.

Technical Description of How the Invention Works

After the bulleted list, proceed into a detailed discussion of how the invention works, accompanied by illustrations. The illustrations should be integrated into the text, and should appear as close as possible to the text that makes reference to them. You may start by pasting in the text of your brief description of the invention, but you'll generally need to edit that heavily. In many cases, it may be easier to start from scratch.

Use some judgment regarding how much of the sundry permutations and options described in your patent application to include. The entire technical brief should be 5-10 pages, so only include elements that are really compelling. Lean heavily on illustrations to make your invention's function clear. The ones in your patent application are a good starting point, but, as you've probably concluded on your own by now, drawings conforming to the patent rules aren't easy on the eyes. They simply take a long time to interpret. Unless your invention is very simple, you will probably want to simplify your images by substituting colored shading for cross-hatching. Absolutely be sure to replace the part numbering with word labels, and leave out labels for obvious unimportant stuff (like fasteners).

Your technical disclosure should include illustrations that allow the gist of the invention to be understood without reading the text.

The reader should not have to refer to the text to understand how your invention works from the figure(s). One clear advantage over your patent drawings is that there's no rule preventing you from combining multiple images of your invention into a single figure, so you can make a step-by-step illustration of how your invention works if it has moving parts or changing states. Show side-by-side images of it folded/unfolded, full/empty, etc.

You'll want to make substantial improvements to any figures you intend to use from your patent application.

Graphics associated with the text should be used sparingly, with the single goal of aiding the reader to digest the information in bite-size pieces. You neither want the reader to feel he/she's reading a novel, nor constantly be interrupted by pointless clip art (really, nix the clip art). Try simultaneously to avoid pages of pure uniform text (without so much as a section heading to break the monotony) while never adding graphics, extra section breaks, etc. that are meaningless for this sole purpose. In short, you should punctuate your text with formatting, but make the effort genuine by using a layout that communicates. This guide is an excellent example (in my most humble opinion).

Use graphics sparingly…

and absolutely no clip art.

Description of the Prototype

Your prototype is key to getting invited to pitch in person – they will want to see it.

Following your description of how the invention works, you should discuss your prototype (absolutely include a photo). In particular emphasize what key development risks you have retired (hopefully all, but certainly this is not entirely feasible in all cases), which will naturally tend to lead you into a discussion of the scope of work required to bring the invention to production. This discussion should be positive, but must be truthful. Spin additional work required as further opportunities for even greater competitive advantages where possible, rather than speaking in terms of what's not tested, but do not lie (apart from the ethics issue, that will come back to bite you when you eventually face open Q&A if you are granted an opportunity to pitch your invention in person). When you address additional development, speak in terms of how many months you project will be required to finish up. Ideally, you'd really like to be able to make a reasonable case that the path to a marketable product is short and low risk (the reason for making a full or partial prototype in the first place). What you don't want to do is make the work ahead sound harder than it will be – when you're marketing, it's better to say the glass is half-full, rather than half-empty (and both would be equally true).

Finishing Up

Close by re-summarizing the benefits provided by your invention.

Finish with a one-paragraph conclusion that summarizes the disclosure and reiterates the advantages of your invention. Contact information should again be included, because some of the reviewers, particularly the technical staff with whom you're trying to encourage early contact, won't necessarily see the cover letter.

In summary, your technical disclosure should comprise:

1. **Contact information**
2. **A one paragraph introduction**
3. **A bulleted list of the advantages of your invention**
4. **An illustrated technical description of how your invention works**
5. **A presentation of your prototype (with a photo) and how it provides for a low risk, short product development cycle**
6. **A summary conclusion restating the advantages**
7. **Contact information**

11.4.3 Video

You can't prevent glitches from acting up in a live presentation, but pre-recorded video's a sure thing.

If a picture is worth a thousand words, a video is worth a thousand pictures. I recommend shooting video footage of your invention doing whatever it does. Nothing will be as compelling as seeing your invention in action, and video provides a much more foolproof way to demo your invention than a live audition. Inevitably, a live demonstration will be expected as part of an in-person presentation, but prototypes do tend to have glitches, and gremlins just love live shows (unfortunately, I'm speaking from personal experience). If you make a video and something goes awry, you can just delete it, fix the glitch, and re-shoot. When it happens live, you can do some combination of stammering explanations and attempting repairs on the fly, but, unless you get it to function, they still won't have seen it work (and seeing is believing). If you've made a video, much (but not all) credibility will be saved.

There's no faster easier way to get their attention than video.

The impulse to click a video attached to an email is almost irresistible (as all virus programmers know).

Of course, a short from that same video can be included in your solicitation, and, if you have it, why not? Now, admittedly email presents some challenges here, since video files tend to be large and many email systems will strip them out of the message, or refuse to deliver the message altogether. Obviously, if your solicitation's traveling snail mail, just including a DVD is one easy way around this. Another, which I have used, is to post the video on a password-protected website to which your solicitation provides a link and a temporary password. These days, that's a lot easier than it may sound to you (and usually costs nothing). Still, if that's difficult (and in many cases it will be), including a video file that is short and compressed enough to be sent electronically without being annoying (<3 MB) is a reasonable option.

Your video doesn't have to look professional to confer a sense of professionalism.

As always, maintain a reasonable level of professionalism. If possible, it's best (although not entirely necessary) to include some narrative in your video. If you do have narrative, make a script, grab a thick marker and posterboard, and make yourself a readerboard. It's OK if your video seems a little home-video-ish – making a professional quality video like the commercials you see on TV takes a lot more than a hand-held video camera, and really would be overkill; but do make sure to eliminate any of the common bloopers (e.g. unplanned cameos by children and pets) that would make your video seem like you didn't make the effort to do your best.

11.5 ORAL BRIEFING

If your invention has real potential and you've applied some quality to your solicitation, wherever it falls on fertile soil you will generally get a response. Your follow-up should be prompt, and should suggest a

Following your solicitation, your next goal is to set up a teleconference.

teleconference to include the responder, and someone from among their ranks who can provide a candid technical evaluation (if that's not the responder). As I've emphasized, making contact and getting buy-in from a technologist who will actually be involved in the development effort is key at this stage. If that goes well and they are genuinely interested, you should expect to be invited to visit them, both in order for them to give you a rundown of their operations/capabilities and a facility tour, and for you to deliver an oral briefing and live demonstration of your invention in person.

Your objective for the teleconference is to be invited to present your invention in person.

Expect to travel, and wear formal business attire.

For your invention to go anywhere, you must expect to travel, and on your own dime. It is not customary for a company to foot the bill for you to come and give your pitch. After all, it's not a job interview, and it's you who is trying to recruit them. When you show up in their lobby, be dressed for success – formal business attire. You never get a second chance to make a first impression. Many people want to believe the world has changed for their convenience, but, in truth, it only has for the 99.9% of the planet who aren't upwardly mobile.

Prepare an electronic slide presentation for your oral briefing.

You should prepare for the oral briefing by creating an electronic slide presentation (e.g. MS Powerpoint). Once again, this doesn't have to be fancy, but it should be comprehensive, and logically organized and formatted for clarity (just a little flare won't hurt, though). I could write another volume on presentation style tips, but here's a very short list that surprisingly few people grasp, and which can really make a world of difference:

- Never include articles such as "a" and "the" on slides.

- Never use font smaller than 14 pt for text intended to be read during the presentation.

- Use a sans serif font like "Arial" or "Geneva" (one without little ticks at the end of strokes like the body text in this guide which is "Times New Roman").

- Avoid carpet slides; that is, slides covered by text of uniform size, spacing, and color. If you use the Microsoft defaults, that's exactly what you'll get – you need to manually adjust the line spacing so different indentation layers stand apart. (I don't think there's a single Microsoft default that's any good – if you use MS Powerpoint much at all and aren't appalled by the default formats, believe me, you need to find someone who can help you with style.)

- Not more than one-in-five slides should be all words with no graphics.

Limit your presentation to 20-30 minutes.

Generally, you should try to keep the length of your presentation to between 20 and 30 minutes, which translates to 10-15 slides. You can include extra back-up slides at the end which you will not use in your main spiel, but which you can jump to as visual aids for anticipated questions.

Confidence is the key to speaking well in front of an audience.

Does speaking in front of an audience scare you? It shouldn't, but, if you're like most, it does. A lot. Such fear is entirely irrational, but that doesn't make it any less real; and you must overcome it. The *Catch-22* of the matter is that the most important ingredient to a good presentation is confidence. This is why there is such a difference between those who speak often and those who do not. Success breeds confidence, and those of us whom life has dealt many an occasion to stand in front of an audience and have fared well grow ever more assured of our talents. Conversely, those who have negative experiences shrink in confidence, and are, therefore, even more fearful at the next opportunity. And so, both success and failure in public speaking are quite self-reinforcing.

Nothing calms the nerves like being well prepared, so practice, practice, practice.

The good news is that whether your fear arises from little experience or bad experience, there is a 100% sure way to overcome it, and that is to practice. I don't mean you have to go to Toastmasters and engage in a personal campaign to become a well-seasoned public speaker. What I do mean is that if you count yourself among the 99% of users of this guide who don't frequently speak in front of an audience, you should practice delivering your oral briefing over-and-over until you can do it in your sleep. That is how good speakers are born – through preparation. Being well-prepared for a specific presentation will bring you the confidence you need to succeed in it; and that success will bolster you for the next. Should such occasions become recurring, you will find that you need less preparation as you accumulate experience. (Some seasoned speakers can stand in front of a random page of the phone book and make a fair go of it; but no matter who you are, presentation is always improved by preparation.)

For the inexperienced, fear peaks just before the start of a presentation, but if well prepared, is gone less than 60 seconds after they begin speaking.

One nearly universal experience for which I would prepare the uninitiated is that, regardless of how prepared you are (remember this fear is, after all, irrational), you will almost certainly feel a knot develop in your stomach as the time for you to stand up and speak draws near, sometimes culminating to sheer terror as the moment arrives. Unfortunately, this is pretty much an inescapable norm for beginners. But I promise you, if you are well-prepared, all you need do is manage to start speaking, and your fear will be completely banished within the first minute, to the point that you will literally be enjoying yourself. For this reason, while there is absolutely no reason for you to rehearse a word-for-word script for your presentation, I do recommend you commit just the first three sentences you intend to say to memory.

Being interrupted by questions is a good sign.

And thereafter make a conscious effort to just relax! You don't have a time limit. Most inexperienced speakers go too fast. Slow down and just talk. Encourage questions mid-presentation, and answer them thoroughly and understandably. If that makes you go longer than 30 minutes, that's OK. It's actually excellent. Questions mean they are engaged and interested – if you're getting questions, that's a good sign. It's when the questions stop that you might want be concerned (but don't sweat it – that's why you're talking to more than one candidate).

Based on the types of questions you receive, you may discern, of course, something about what the participants are thinking; but, I can pretty much tell you in advance (it's really not very difficult – just ask yourself what you would be thinking):

> **What potential licensees will be thinking as you present your slides:**
>
> - Who are you? Are you special in some way?
> - Do you really own the invention/Do we need you? (Is it truly novel?)
> - Can your invention provide an edge over the competition?
> - What's the risk of ignoring it?
> - Can we make it work?
> - Can this be profitable/will people buy this?

Your presentation should touch on all of the above questions.

You should, as you might expect, start your presentation by introducing yourself along the lines of what you do for a living and how you became interested/involved in the subject matter of the invention. You don't need a slide for this; just tell them. As you work your slides, let the above questions focus your content.

Take pride in the fact that you're really doing something most people would admire.

Be confident and enjoy the experience.

Your general demeanor should be direct and friendly. Inexperience can tend to make for nervousness, but will it help for me to assure you that there's no need to feel that way? I expect no, but the honest truth is that you should enjoy yourself. You're doing something exciting, meeting with interesting people, and you're the real deal. The people you want to connect with go to work the same way you (hopefully) do, aiming to respect and get along with the people they work with, and they will gladly treat you the same way. You aren't there to defend yourself and you're not wasting anyone's time. You aren't there to beg. You're making them an offer that you regard as a good opportunity. Above all else, the most key element of professionalism in this context is a relaxed, friendly, and patient disposition. For them it's just another day at work, and your presentation is probably the most interesting thing in which they'll participate all day. Smile.

Make sure your prototype is with you and functioning.

Also, if your prototype is at all portable, you must bring it with you ready to perform. I'm sure you can appreciate that its absence would be regarded as quite suspicious.

11.6 Non-Disclosure Agreements

Don't expect a prospective licensee to sign an NDA for an invention sight unseen – that's why you filed for a patent in the first place.

Most people are familiar with the concept of a "non-disclosure agreement", or "NDA", which is a contract entered into by one party promising not to disclose information provided by a second party for a set period of time (since never is a very long time, most expire after three to five years). Most inventors expect to need to have prospective licensees sign an NDA before they feel safe to disclose any details of their invention to them. This expectation is false, exactly because you have filed a patent application. You filed that patent application before marketing for the simple reason that no-one would be willing to sign an NDA until they have seen and understood your invention. Until they know what it is, they can't be sure they aren't agreeing that you own rights to something they already have in the works, or, at least, overlaps or has common elements with said thing they already have in the works. So, if you request a prospective licensee sign and NDA without having seen anything, you'll get nowhere fast and just look naïve (generally, no-one will hold that strongly against you – a degree of naiveté is perhaps part of the expected charm of a free spirit like an independent inventor, but the point of this guide is that you be informed).

Generally an NDA will come into the picture as part of a license agreement – to keep you quiet, not vice-versa.

You won't generally need an NDA after you've shown them your disclosure either, as your patent packs all the defense you really require, and extra publicity can only work in your favor. If they really think your invention has potential, they'll naturally want to keep it to themselves. A non-disclosure clause will normally be included in a licensing agreement, but that will be more to keep you quiet than vice-versa.

11.7 Negotiations

Consider the affect of already having other auditions scheduled on your ability to negotiate.

If you know enough about your candidates to prefer one, it's best to go into that meeting in a position to deal.

If your oral presentation goes well, they may simply state that they are impressed and will be in touch. That may be convenient if you already have other auditions lined up, because once you schedule them, you really can't make an exclusive deal with anyone until you follow through without burning bridges. Promising as a preferred prospect seems, it could still fall through, so you'll want to keep your options open (and who wants to be a rude jerk, anyway). Alternatively, after begging your patience to give them 20-60 minutes to talk amongst themselves, they may want to negotiate immediately. (Think about those dynamics when you make decisions about whether or not to queue-up multiple auditions, and in what order to schedule them.) If you have other interviews scheduled, even if you have reason to really prefer the candidate in front of you, you should be honest about that fact and that you feel you must honor your previous agreements and give the others their fair chance. If they're very attracted to your technology, they may find this a little frustrating, but they'll respect your integrity. Conversely, if the offer really is something you want to jump on, you can simply politely cancel your other engagements (business is business, after all), but, as I said, be aware you're sacrificing back-up options.

Wait for the prospective licensee to broach the subject of compensation.

If you cannot immediately commit, there's no harm in stating that you consider them to be a leading candidate, if that's true based on what you know about them and what they've shown you (but be prepared to be specific about why you feel that way). There's also nothing stopping you from discussing the terms of a potential agreement, if they turn the conversation in that direction, you just won't be able to sign on the bottom line. You shouldn't be the one to bring up compensation – that's their job, and everyone knows it must eventually come to that. Generally speaking, they commonly will not want to broach the topic of royalties (other than just basic sounding to make sure you're prepared to be reasonable) until you have already expressed that you regard them with favored status.

Never speak about one potential licensee to another.

Never speak to one prospective licensee about another by name. That's very bad form. You can subtly (preferably indirectly) refer to other prospects to convey a certain sense of urgency and credibility, but that is all. You really should avoid talking about the competition completely, if at all possible. If you are asked to whom else you are talking, just state, in a positive and polite way, that you must respect other potential licensee's privacy, just as you will theirs; but assure them that the process is not a matter of competition. They know you're evaluating others; and it's certainly not in your best interest to pretend you're not; and they know there's an element of competition involved; but you should always emphasize your seeking the right "fit" for your invention.

Do not let your marketing strategy take the appearance of an auction.

In the modern politically correct world, it's uncouth to imply that any organization or anyone is better than anyone else – they're just "differently abled." Now, I can't say that my experience supports that pleasing notion, even though I really like the sound of it. Near as I can tell, some people are more talented than others, and I'll refer you to a higher power if you have a need to understand where the fairness lies in that. Likewise, some organizations are just going to have it together much more than others, which reflects directly on the competency of the management and workforce; but, we don't need to be so impolite as to openly talk about it. For the moment, just note that as you dialogue with multiple candidates, as an outsider it's somewhat inappropriate to talk in terms of superiority/inferiority, and potential licensee's are going to be very turned off if they feel your process is a straight-out auction. Don't present the situation as a bidding war, and it really should not be. Just because one candidate offers you a higher percentage cut doesn't mean they'll provide you more revenue. Your final evaluation should not be based on what the licensee offers, but your assessment of what they can actually deliver. Would you rather get 5% of $500k/yr or 10% of $100k/yr?

There are a number of fairly standard topics that one would expect to find in a licensing contract, here listed to provide you general awareness:

Assignment	Typically an exclusive licensing agreement will assign ownership of (and the responsibility to prosecute) the patent to the licensee in exchange for royalties.
Indemnity	It is important and absolutely standard that the licensee hold harmless or "indemnify" the licensee. Basically, if they market a product and somebody gets hurt by it and sues, you are immune to prosecution. The licensee takes full and sole responsibility to ensure fielded products are safe.
Successorship	Royalties are to be treated as inheritable property, and the agreement should state as much. From a pragmatic, if not melodramatic standpoint, if your invention turns out to be worth millions, no-one can avoid payment by taking you out – they'd still have to pay your death beneficiaries.
Termination Clauses	An agreement may set conditions for voluntary termination by either party, but typically such clauses are for the licensee.
Breach Remedies	Typically an agreement will spell out what happens if someone transgresses its terms. Often this will include an agreement to settle disputes through arbitration.
Non-Disclosure	As mentioned above, most license agreements incorporate mutual non-disclosure.
Royalty	Obviously somewhere must be a statement of what the licensee agrees to pay you. We'll discuss what to expect in Section 11.9.
Minimum Royalty	In Section 11.9 we'll also come to why any agreement you enter into should always specify a minimum royalty, which is a minimum payment allowed per payment period, irrespective of gross sales volume.
Sublicensing	Typically, a license agreement will provide the licensee the right to sublicense the invention to others with a specified agreement on what the licensee receives from such agreements (either a fixed royalty, or a percentage (half seems standard) of sublicense royalties).

Any license agreement will also contain sundry administrative elements, such as party identifications, definitions, warrantees, statement of within what state's laws the contract is defined, official correspondence rules, etc. It's a good idea to have a legal expert review an agreement before you sign on the dotted line (especially if you're not fairly fluent in legalese).

Of course, it's always best to have a legal expert review any contract you plan to sign.

A license agreement will take place in two stages. First will come a verbal agreement, which will be where you sit down, discuss, and settle on basic terms of the contract, which should include a royalty and minimum royalty. Do not think that a verbal agreement is non-binding! Once you make a verbal agreement the law does provide both parties some remedy if the other backs out without good cause.

Verbal agreements confer obligations.

That verbal agreement will be followed by a written agreement, which will set forth all of the miscellaneous details that are impractical to flesh out in open discussion, and many of which are part of the company's standard boiler plate. You can back out of the agreement if there are terms that you find unacceptable, but you should not. Instead, simply object to those terms and make a counter proposal. They may accept your counter, or counter again. Be reasonable and cut to the chase – too much back-and-forth can really sour a relationship. Remember, most of the terms in such a contract will be regarded by them as standard, and, if they are people you really should be working with, fair. There shouldn't be much need for dickering unless you've fallen in with wolves. If you really can't live with their terms, ultimately you must provide a best-and-final offer (identified as such), which means you walk away from the table if they do not agree. That's your way out, if you detect that you're being swindled in the language of the proposed contract.

Do not attempt to dismiss a prospective licensee with whom you have a verbal agreement – instead, if you find the proposed terms unacceptable, make a counter offer.

If differences are irreconcilable, the correct exit strategy is to make a best-and-final offer.

11.8 EXCLUSIVE VS. NON-EXCLUSIVE LICENSING

In the end, if (hopefully) your invention really is a winner, more than one licensee will likely want it. You can deal with everyone honestly and with integrity, but you won't be able to make them all happy, because they have conflicting interests. So why not go with multiple licensees? In a nutshell, it's extremely unlikely that any licensee will be willing to sign a non-exclusive agreement. You're asking them to foot the bill for quite a bit of development work and marketing in exchange for uncertain returns. (I'm sure your invention is great, but in the world of inventions, nothing comes close to qualifying as a sure thing.) Who would take all that risk and expense only to let others copy them (after all, aren't you the one who just finished writing a patent application)? Non-exclusive licenses are common, but they apply to already developed products. If you have completed the development of your invention, established a production line, and made a noticeable splash in the marketplace, you're in a position to offer non-exclusive licensing. If not, a non-exclusive agreement has little to offer a licensee.

Non-exclusive licenses really only apply to already successful products.

From a marketing standpoint, exclusive licenses are offence, while non-exclusive are defense.

Exclusive and non-exclusive licenses are really very different animals from the licensee's perspective. An exclusive license represents a genuine asset providing an edge to hopefully out-compete rivals. A non-exclusive license is more like an entry fee to be allowed to compete at all, providing only access to an otherwise level playing field.

To the manufacturer, a non-exclusive license is much like a tax.

Let's consider for a moment you've invented something that brings novelty to a well-established field where the well-over-twenty-year-old incumbent technology is pushed through brute-force marketing power amidst rivals lacking any substantial technical advantage over one another. Clearly exclusive access to a superior technology would be very attractive. But, what about a non-exclusive license? You'd be effectively offering nothing more than to become a monkey on their backs. If you licensed your invention to one company non-exclusively, you'd very quickly end up with licenses to all of their competitors as well. That'd be great for you, but the licensees would all find themselves in the same market gridlock they were before, and now having to pay you to be there! So you see, non-exclusive licenses are only entered into as a necessary evil. Even for established products that come with practically guaranteed market share, prospective partners push for regionally exclusive licenses, which provide them sole distributorship over a specific territory.

From a business perspective, it's not what's sold, but who's selling it that counts.

From a purely business perspective, companies don't care in particular what they sell, as long as they're the one selling it. Sure among their ranks there typically exist those vital individuals with a true appreciation and enthusiasm for the technology, but they are merely a means to an ends (making money). A perfectly thriving transportation industry existed long before the invention of cars, and people who sold horse carriages weren't happy for the change.

Non-exclusive licensing of a not-yet-fielded invention will effectively render the patent worthless.

So, if you approach a company with an offer for exclusive licensing, that's a win-win business proposition; but if you offer a non-exclusive license, you're effectively taking the role of an extortionist. They'll pay an extortionist if they have to survive (if you can't beat 'em, join 'em, and all that), but they'll look for a way out. Anyone who did sign a non-exclusive license with you for your as-yet-not-fully-developed invention would almost certainly only do so if it required no minimum royalty payment (which you should never agree to under any circumstances, but we'll come to that). I'm not suggesting that you're likely to encounter anyone who'd attempt such a thing, but so doing would be a nasty trick, as they'd hold a license to practice your invention, but do nothing to develop it. They'd simply have hedged their bets for free, and you would hence lack the ability to exclusively license your invention anywhere else (which pretty much encompasses its only true value, so your patent would now be worthless).

Bottom line – you're in the market for an exclusive license.

To sum up, when you approach a prospective licensee with a brand new product, it essentially goes without saying that what's on the table is an exclusive license.

11.9 COMPENSATION

Royalties should be based on net sales.

If all goes well, eventually must come the ugly business of setting the licensing price. Licenses typically yield revenue known as a "royalty", which is a percentage payment based on sales. Accept no royalty that is based on any number other than net sales. It may seem logical and more fair to base a royalty on net profit, so that when the licensee makes money, you make money; but, let me assure you that is absolutely not the case. Many considerations play into how much profit is attached to a product, and that value can change frequently. As part of a marketing strategy, your product could be sold temporarily at little or no profit. If your royalty were based on profit only, you'd proportionally make little or nothing, but a royalty based on net sales (the total proceeds) will still yield a respectable return. A licensee needs to be free to alter price without disproportionately impacting your income per unit sold. Also, profit (net sales minus cost) can be quite subjective, because what rolls into cost can be open to interpretation. Does marketing count as cost? Management costs? How about a portion of the electricity bill? Net sales is cut and dry – gross sales minus product returns, and so royalties based on net sales are standard.

What constitutes a fair royalty depends upon a number of factors.

Of course, the most burning question in your mind at this point will inevitably be: What's fair?. You could be throwing an opportunity away if you expect too much, and sell yourself short if you expect too little. Unfortunately there is no easy answer, because a number of very case-specific factors must be considered, such as:

- **Volume of expected sales**
- **Profit margin**
- **Competitiveness of the industry and product area**
- **Maturity of the invention**
- **Development risk (remaining technical issues affecting if the invention can be converted to a practical product, how much it will cost, and how long it will take)**
- **Production set-up costs and capital requirements**
- **Significance of the competitive edge provided by the invention**
- **Stability and history of the licensee**
- **Resources that the licensee brings to the table**
- **Presence or lack of alternative products in the marketplace**
- **Growth or recession rate of the relevant industry**
- **Market risks associated with the invention**
- **Safety-related liabilities associated with using the invention**

What constitutes a fair royalty is genuinely difficult to determine.

What prospective licensees will think is fair may be even more mysterious.

Most royalties fall into the narrower range of 4 to 9%

An unusually low or high profit margin is the most likely basis for a royalty outside the typical range.

One thing that you can, at least, evaluate objectively is to know what would be the lowest royalty worth having at all.

The list goes on. Given the large number of factors, it should not surprise you to learn that royalties typically earn anywhere from one to twenty percent. Clearly, one could hypothetically create weighting indexes for all of the factors listed above, and others, and write a formula that outputs the correct, fair royalty for any invention, and maybe someday someone will publish just such a thing. Naturally such an equation would be useful only if it had general acceptance by licensors and licensees as a standard guideline; but even still, note how difficult the above factors are to quantify. And then there's a wildcard we could call the human factor, which isn't based in rationality, is entirely unpredictable, and may relate more to what people ate for breakfast that morning than anything else.

So how can you estimate what's fair for your invention? If you scan the internet you'll find a number of websites offering to sell you access to databases listing stats for various license agreements for the low price of several thousand dollars a year and up. I've never subscribed to one of those, and expect that if I did, I'd find lot's of data illustrating that licensing agreements span a range of 1-20% and that a large number of hazy factors seem to influence the royalty in predictable but difficult to quantify ways, except for some that made no sense at all. (I'd feel pretty silly if I paid $4k for that.) One thing I can offer is that, most commonly, royalties fall between the narrower range of 4-9%, and, unless you have a particular reason to expect otherwise, plan to negotiate in that range. You have genuine reason for concern if a licensee tries to convince you to accept an offer below that range (but for some inventions a lower value really may be fair), and you may need a reality check if you think you're going to walk away with more.

Profit margin will generally be the only factor likely to form a legitimate basis for a royalty out of the common range. This is because the royalty is computed as a percent of net sales, not profit. It would be quite unrealistic to expect to collect 5% on a product selling at 9% profit margin, for instance. You'd be making more than the licensee! That might make you happy, but don't expect a licensee to go for it. After all, they are carrying all the risks. Conversely, in cases where the profit margin is unusually high (like some biomedical products), it is reasonable to expect higher than normal royalty percentages.

There is one benchmark you should walk into negotiations knowing, and that is what royalty you need as a minimum to make the effort worth your while. Forget about the time and effort you put into the prototype, and the hours spent on your patent application. That's all water under the bridge. Going forward, you need to understand what minimum income yield you regard as worthwhile, below which you'd basically rather call it quits and move on. Take that number in $/year, and divide by the revenue estimate you computed back in Chapter 2. Generally, if you made the correct call when you evaluated whether or not to move forward with your invention, that percentage will be pretty low, but it's a useful metric to keep in the back of your mind.

Make your royalty a good deal for both you and your licensee.

A good royalty isn't only about maximizing your cut. Of course you want to make the most money you can, but recognize that maximizing your royalty won't necessarily accomplish this any more than raising the price of product guarantees increased revenue. (Recall from economics that an optimum price exists where the product of profit-per-unit and units sold (= revenue) reaches a maximum. Sadly, most people don't get any economics in their education anymore, going through life without which is similar to sky-diving without a clue about what gravity does). In the end, you want your licensee also to be getting a good deal, as this will make your project compete better against their other development efforts for attention and resources. If you're smart, you'll balance their interests and yours.

Still, at the end of the day, it's fair to expect a high degree of interest to translate into a better offer.

While increasing your royalty won't necessarily boost your profit, it is fair, taking into account the other factors, to regard the level of offered royalty as some reflection of the motivation and anticipated returns of a potential licensee. Obviously, a prospective licensee who brings more to the table would expect to be able to win your business at a lower royalty to one who has less promise, so one cannot compare numbers directly. Nevertheless, if a licensee is really convinced that your invention is a must have, that they don't want to risk you're going with someone else can be expected to be evident in the offer.

Choose your licensee carefully. If they fail, you'll not likely get a second shot.

Choose your licensee carefully. Sadly, if you sign with a licensee that drops the ball, chances are the dream's over. When you select a licensee, you really are throwing your lot in with them, for better or worse. Of course, you can claim damages if they do something really inexcusable, like allow your patent to lapse; but generally, if they don't perform well, you'll be hard-pressed to find a way to take it back.

License agreements will typically include some minimum performance requirement, which can take one or both of two forms:

The contract can require the licensee to keep pace with schedule milestones leading to the product to be fielded by a specified date.  The contract can specify a minimum royalty payment per period.

Few licensees will be willing to sign an agreement that forfeits all rights if they fail to perform, in anticipation that delays may arise from circumstances not entirely under their control. As such, I strongly recommend you insist on a minimum royalty.

> **Never sign a licensing agreement without a minimum royalty.**

A minimum royalty provides an ongoing incentive for the licensee to perform.

A minimum royalty won't be enough to make you rich, but it will ensure that the licensee has honest intentions plus will provide a constantly renewed incentive to perform (no licensee will like paying a royalty on something that is generating no revenue). To meet intent, the minimum royalty shouldn't be (and won't be) an enormous cost to the licensee, but it shouldn't be the kind of money they'd pass by on the sidewalk either. Also, if the project doesn't go anywhere, a minimum royalty will guarantee at least some return for all your efforts, and that's certainly much better than nothing.

11.10 INFRINGEMENT

Usually companies will recognize it's in their best interest to honor valid patents.

If a company ignores your patent, it generally means they think they have legitimate grounds to do so.

Of course, you may continue to worry that one of the companies to whom you send your technical disclosure will steal your invention. There is, of course, that possibility, though it is actually fairly unlikely unless they don't agree that your invention meets the requirements to qualify to receive a patent. Certainly if they're copying it, they recognize its utility; but they may have reason to believe that it isn't novel, or that it is obvious. One thing you can do to minimize the chance of this happening is to always be reasonable, i.e. make the trade between the cost/benefit/risk analysis favor working with you rather than around you. But still, for whatever reason, we've all heard those stories of inventors who complain of being steamrolled by a corrupt giant. If you do see a company infringing on your patent, there is one cardinal rule of which I cannot maintain a clear conscience and neglect to mention:

> **Never warn someone that you believe they may be infringing on your patent.**

Notifying a company that you think they are infringing gives them grounds to head over to argue to have your claims dismissed in a local court without even telling you.

Why? Because laws are basically made by lawyers for lawyers, and in the world of patents some of them have succeeded splendidly in making it effectively impossible for you to be civil instead of taking civil action. How? By passing laws that allow the alleged infringer to go to a local kangaroo court and get a summary judgment against you without even notifying you. They can't take such action, however, unless you raise the issue of infringement first (say, by sending them a letter).

You've probably already surmised why this works out well for lawyers – since you can't even open an honest dialogue with an infringer (who, in many cases may not even be aware that they are infringing) without

The law is pretty much set up so that you and an infringer can't even try to work things out outside of court.

inviting disaster, what you're left with is simply no option other than to make sure the first they hear of the matter is that you are suing them (Ch-ching!). Job security doesn't get much better than that.

If you think someone is infringing upon your patent, the first thing you should do is examine their product as closely and objectively as possible. Have they genuinely infringed? Remember, there are two possibilities here. Either:

They are aware of your patent.

If they know about your patent (say, perhaps you already sent them a solicitation), then clearly they think they have grounds to show that either they aren't infringing, or that your patent is invalid. There is the third possibility that they simply don't think you will find out about what they are doing, or at least won't take any action.

or

They are not aware of your patent.

If they are oblivious to your patent (after all, there are a lot of patents out there), they may really be innocently infringing, but, that doesn't mean they haven't done anything wrong. Be careful, though – if push comes to shove, they may be able to show that they developed the IP first.

Remember, if a company thinks your patent isn't valid, it's not their job to challenge you, it'll fall to you to challenge them.

Recall that if a company wants to market a product that would infringe, but doesn't regard your patent as valid, they won't notify you or outright challenge your patent. They'll just move ahead with their plans and wait to see if you to come after them. (It would certainly reduce their financial risks to challenge you at the outset, but it weakens their position since so doing lends credibility and relevance to the patent.) As the patent holder, you're the infringement police. Of course, if you've already licensed your invention, no additional action is needed other than to sick your licensee on them (assuming it's an exclusive agreement).

The only safe way to fire a warning shot across the bow is to send a solicitation to license which makes no mention that you believe they are already practicing your invention.

If you haven't already licensed, before you log into scummylawyers.com to find a legal pit bull (more seriously, we've already been through how to find a decent patent attorney in Chapter 9), there is one way you can both warn an infringer and strengthen any potential case you might have against them at the same time. If you haven't already done so, one option is to send them a solicitation to license. Such an offer should make absolutely no mention of the fact that you believe they may be infringing, just give them a chance to acknowledge interest in licensing. As before, sending them a solicitation just opens the door for communication – it does not obligate you to license to them.

Of course, upon receiving your offer, they will make a cost/benefit/risk assessment of whether the best course of action is to diffuse the situation by pursuing a license agreement and/or desist in their infringing

Upon receiving your solicitation, an infringer will make a cost/benefit/risk assessment about whether to cooperate or fight.

activities, or prepare for a fight. The moment they respond with some interest you'll have them in the noose (and they know it), so they'll be cautious. Unlike an ordinary cold-contact solicitation, if you really would prefer to license to the infringer (and I urge caution unless you are fairly certain their infringement was accidental), it may be constructive to propose terms (royalties) in generic language in your cover letter to provide them what they need to make an informed decision regarding their course of action (since no-one is in any position to talk to one another).

If a solicitation evokes any response making claim to any infringing intellectual property, you must either immediately file suit or fold.

They may counter with fighting words to the effect of "No thanks, we already have this" or "We believe you are infringing on our intellectual property"; translation: "Get lost!" The moment you receive any communication suggesting any dispute of intellectual property, the glove has been thrown down, and you have only two choices – fold or take immediate legal action. Once the dispute is, in fact, a dispute, they may head off to kangaroo court, as may you. I'm sure it won't surprise you that whoever files suit first has an edge.

As a plaintiff in a land governed by deep pockets law, at least you can pursue the case on contingency (assuming you can stand the smell).

If the infringer's enterprise is large at all, however, you do have one irrevocable advantage, and that, naturally, is that there's a good chance you'll be able to get your lawyer to take the case on contingency. As always, until you have actual revenue, you should religiously minimize your out-of-pocket expenses. That such practices are completely unchecked and rampantly pillaging society-at-large may give you pause, but paying out-of-pocket for a capricious verdict hardly seems a fair alternative. Be assured there's a ~50% chance you'll lose the suit even if you turn out to be in the right. (Common sense and fairness seem to be reserved for small claims court on television.)

Litigation is for the professionals; all you really need to know is that any action taken in response to infringement by you or your licensee must be swift and decisive.

But, in this final discussion we digress. Generally, most people and organizations will have the sense to do almost anything to stay out of court (just as they avoid shark-infested waters when going for a swim.) The main point here is that if fortune/fate deals you the ugly business of addressing infringement, the law has been manipulated so as to effectively require you to strike swiftly and precipitously, and I would be remiss to not alert you to this fact. Beyond that I can only recommend you seek the advice of a good attorney (preferably the one who reviewed your CLAIMS per the recommendations of Chapter 9), but at least you now know that you need to do so expediently.

11.11 OFF YOU GO.

It's a long road ahead, but you're on the right track.

And so, dear inventor, you have successfully completed your patent and are now off to market the fruits of your labor. My hat's off to you, and what I hope to be a bright future. Very few people get this far, and now you have. That this guide should finish in Chapter 11 is perhaps ironic, as great emphasis has been placed throughout on helping you to avoid exactly that fate. Remember that a whole lot of things have to go right to make an

May this be one of many accomplishments for you.

invention profitable, but only one has to go wrong to make the whole thing a bust. Still, let me assure you that no matter how things turn out, your time and effort have not been wasted. Accomplishing such an undertaking has value in-and-of-itself, and you are now officially a member of the select few who do more with their dreams than procrastinate. If only such endeavors were habit-forming!...but unfortunately, if you're like most, you'll have to continue to work to keep up the habit. Do that. Create, and never stop creating. I can't make any guarantees as to what lies ahead, but I can assure you you'll be proud of what's behind.

Until next time,

R. K.

P.S. There is something you can do for me. Please share with me your experiences, both good and bad. As you have benefited from those of others through this guide, so should future users benefit from yours. I cannot respond to every email, but you have my attention:

http://www.inventionpatentinformation.com

Index

27B|6 (27B-stroke-6), 138
Abstract of the Disclosure, 131–32
Action steps
    1. Profitability assessment, 20
    2. Patentability assessment, 21
    3. Image viewer download, 24
    4. Patent search, 31
    5. Prototype fabrication, 40
    6. Drawings, 67
    7. Patent title, 69
    8. Inventors list, 70
    9. Field of the invention, 72
    10. Introduction, 74
    11. List of remedied problems, 75
    12. Brief summary of the invention, 78
    13. Drawing descriptions, 80
    14. Detailed description of the invention, 88
    15 Inserting reference numbers in text, 92
    16 Inserting reference numbers in drawings, 93
    17. Closing statement, 95
    18. Claims, 127
    19. Abstract of the disclosure, 131–32
    20. Form PTO/SB/05, 136
    21. Form PTO/SB/17, 141
    22. Form PTO/SB/01, 143
    23. Document receipt postcard, 148
Adobe Acrobat, 132
AlternaTIFF, 23
American Inventor, 6
Amino acid sequence listing, 143
Application Data Sheets. *See* Forms
Application, patent
    elements of, 44, 45–46, 132
    forms. *See* Forms, application
    general requirements, 44–45
    mail to address, 146
    needed tools/skills, 41–44
Assignment, 174
Atomic Energy Act of 1954, 17
Attorney, patent, 7–9, 107
    and infringement, 180–82
    and patent language development, 115
    and priority date transfer, 142
    and skip numbering, 91, 92
    claims review, 128–30
    commercial setting, 8
    engaging, 8
    power of, 134
    timing for involvement, 8
Background of the Invention, 71–75
    Brief Description of the Several Views of the Drawing, 78–80
    Brief Summary of the Invention, 76–78
    Description of the Prior Art, 72–75
    Field of the Invention, 71–72
Breach remedies, contract, 174
Claims, 96–130
    adjectives, use of, 109–13
    and preferred embodiment, 120
    and trademarks, 115–17
    apparatus claims, 120–22
    avoiding absolutes, 113
    avoiding use of alternatives, 113
    claiming the result, 117–18
    connecting words, 108–9
    dependent, 97–98
    design matrix, 126–27
    Examination Support Document, 99–100
    means-plus-function language, 114–15
    method claims, 120–22
    multiple dependent, 103–4
    number limit, 99
    professional review of, 128–30
    proficiency self test, 122–26
    purpose, 96
    qualifying language, 105–7
    strategy, 96–100
    structure, 100–104
    unnecessary supporting elements, 119–20
    using consistent terminology, 107–8
Commissioner for Patents, 146
Compensation, 177–80, *See also* Profitability
    royalties. *See* Royalties
Continuation, 9, 110, 135
Dependent claims. *See* Claims
Design patent. *See* Patent: types
Detailed Description of the Invention, 81–94
    and claims, 81–82
    Closing Statement, 94
    part numbering, 82–83
    reference coordinates, 85–86
    reference numbers. *See* Reference numbers
    vocabulary, 83–85
    writing procedure, 86–88
Divisional patent application, 99
Drawings, 46–67
    adding dynamic reference numbers, 92–93
    black-and-white preferred, 50
    cross-hatching, 43, 52–53
    enlarged views, 51
    examples, 48–49
    importing, 66–67
    proofing, 65
    quality, 65–66
    rule enforcement, 47–48, 64

sectional views, 51
separation of, 50–51
software, 42–43
text, 51–52
USPTO standards, 53–64
Ebay, 23
Edison, Thomas, 67, 70
EFS-Web, 143–46
   and fee transmittal form, 139
   and transmittal form, 133
Electronic filing, 143–46
Evaluating (an invention), 6–7, 17
   profitability, 18–20
Examination Support Document, 99–100
Fees, 137–39
   and small entity status, 136–37
   and USPTO, 2
   Fee Transmittal Form PTO/SB/17, 136–41
   international patent, 5
   late, 138
   maintenance, 138–39
FIG. *See* Drawings
Forms
   Application Data Sheets, 141
   Declaration for Utility or Design Patent Application Form PTO/SB/01, 141–43
   Fee Transmittal Form PTO/SB/17, 136–41
   Utility Patent Application Transmittal Form PTO/SB/05, 133–36
Forms, application, 132–43
Franklin, Benjamin, 95, 125
Freepatentsonline.com, 24
Google Patents, 29
Indemnity, 174
Infringement, 180–82
Inkscape, 42
Intellectual property, 1
   and national prosperity, 2
   as an invention, 2
   attorney, 8
   Brown & Michaels Intellectual Property Home Page, 149
   defined by claims, 37, 81
   in proprietary notice, 165
   infringement, 182
   not defined by Detailed Description of the Invention, 81
   World Intellectual Property Organization, 5
InterneTIFF, 23
Inventor's notebook, 10–11
ITBAT, 84–85
   and final text review/editing, 88

and means-plus-function language, 114–15
   examples, 85
Leahy-Smith America Invents Act, 15, 146
   micro entity, 137
   non-electronic filing surcharge, 143
License agreement. *See* Licensing
Licensee selection criteria. *See* Licensee selection criteria
   competing market share, 151–52
   heritage, 152–54
   relevant knowledge, 154–55
   size, 151
Licensing, 150–82
   exclusive vs. non-exclusive, 175–76
   finding prospective licensees, 155–56
   license agreement, 174–75
   making contact, 156–59
   soliciting prospective licensees. *See* Solicitation
   sublicensing, 174
Memorandum of Invention, 11
Micro-entity, 137
Microsoft Powerpoint, 169
Microsoft Word, 42
   adding reference numbers. *See* Reference numbers
   importing drawings, 66–67
Multiple dependent claims. *See* Claims
NDA. *See* Non-disclosure agreements
Negotiations, 172–75
Non-disclosure agreements, 172, 174
Non-obviousness. *See* Patentability
Novelty. *See* Patentability
Nucleotide sequence listing, 143
Oral briefing, 168–72
   slide presentation, 169
   speaking before an audience, 170–71
   travel, 169
Patent
   continuation. *See* Continuation
   enforcement, 2–3
   frequency, 13
   provisional, 3–4, 10
   types, 3–4
   why issued, 1
Patent Cooperation Treaty, 5
Patent search, 22–33
   by keyword, 24–27, 31–32
   by US Classification Code, 27–29, 32
Patent writer, professional, 7
Patentability, 13–16
   non-obviousness, 16
   novelty, 14–15

usefulness, 15–16
PDF writers, 145
Plant patent. *See* Patent: types
Post card, document receipt, 146–48
Preliminary invention protection, 10–11
    inventor's notebook. *See* Inventor's notebook
    Memorandum of Invention. *See* Memorandum of invention
Priority date, 9
    and continuations, 135
    and document receipt postcard, 146–47
    and provisional patents, 3–4
    documents, inclusion of, 135
    transfer, 142
Profitability
    evaluating, 18–20
    vs. risk, 5–6, 9, 5–6
Promoters, 11–12
Prototyping, 12, 34–40
    design, 37–38
    development considerations, 35–36
    partnering, 39
    strategies, 36–37
Provisional utility patent. *See also* Patent: provisional, types
Reference numbers, 89–94
    adding dynamic numbering to drawings, 92–93
    adding dynamic numbering to text, 89–91
    skip numbering, 91–92
Resources, other, 148–49
Restriction requirement, 99
Royalties, 174
    assessing fair value of, 178–79
    basis of, 177
    minimum, 174, 179–80
Search engine, internet
    and finding gov't documents, 132–33
    and finding licensees, 155, *See also* Licensing
    Google, 23, 149
    USPTO patent database, 23, *See also* Patent: search
    Yahoo, 23
Small entity status, 136–37
Solicitation (of prospective licensees), 159–68
    and compensation, 160
    cover letter, 160–63
    strategy, 159–60
    technical disclosure. *See* Technical disclosure
    video, 168
Specification, 68–130
    Background of the Invention. *See* Background of the Invention
    claims. *See* Claims
    Detailed Description of the Invention. *See* Detailed Description of the Invention
    elements of, 46
    inventors, list of, 70
    title, 68–69
Sublicensing, 174
Successorship, 174
Technical disclosure, 163–67
    cover sheet, 164–65
    description of how the invention works, 166–67
    introduction, 165–66
    proprietary statement, 164–65
    prototype description, 167
Termination clauses, contract, 174
Timing, 8–10
    filing before licensing, 9
    risk, 9–10
    when to involve an attorney, 8
Trademarks, 115–17
United Nations, 5
United States Patent and Trademark Office, 2
    abstract of the disclosure, 131
    and promoter warnings, 11
    application elements, 132
    claim format preferences, 101
    copyrights, 74
    examiners, 65
    fees. *See* Fees
    image viewers, 23
    links to international patent information, 5
    nucleotide or amino acid sequences listings, 143
    office actions, 9
    patent protection, 2
    patent requirements, 42, 45
    patent searching, 6, 23–25, 27, 31, 72
    patent types granted by, 3
    priority date, 9
    provisional patents, 3, 4
    regulation change notification, 107
    requirements, 71, 84, 92
    resources for the inventor, 7
    self-sufficiency, 2
US Classification Code, 27–28
Usefulness. *See* Patentability
USPTO. *See* United States Patent and Trademark Office
    drawing standards, 53–64
Utility patent. *See* Patent: types
Washington, George, 13
WIPO. *See* World Intellectual Property Organization
World Intellectual Property Organization, 5

Made in the USA
Lexington, KY
27 December 2011